Macromonomers as a Synthetic Tool for the Construction of Complex Polymer Architectures

Zur Erlangung des akademischen Grades eines

DOKTORS DER NATURWISSENSCHAFTEN

(Dr. rer. nat.)

Fakultät für Chemie und Biowissenschaften

Karlsruher Institut für Technologie (KIT) – Universitätsbereich

genehmigte

DISSERTATION

von

Anna-Marie Zorn

aus

Sinsheim

Dekan: Prof. Dr. Martin Bastmeyer

Referent: Prof. Dr. Christopher Barner-Kowollik

Korreferent: Prof. Dr. Michael A. R. Meier

Tag der mündlichen Prüfung: 20.04.2012

Bibliografische Information der Deutschen Nationalbibliothek

Die Deutsche Nationalbibliothek verzeichnet diese Publikation in der Deutschen Nationalbibliografie; detaillierte bibliografische Daten sind im Internet über http://dnb.d-nb.de abrufbar.

ISBN 978-3-8325-3197-3

Logos Verlag Berlin GmbH
Comeniushof, Gubener Str. 47,
10243 Berlin
Tel.: +49 (0)30 42 85 10 90
Fax: +49 (0)30 42 85 10 92
INTERNET: http://www.logos-verlag.de

Die vorliegende Arbeit wurde von Juni 2009 bis März 2012 unter Anleitung von Prof. Dr. Christopher Barner-Kowollik am Karlsruher Institut für Technologie (KIT) – Universitätsbereich angefertigt.

Abstract

Development and innovation in macromolecular science focuses on the improvement of existing synthetic protocols to obtain well-defined polymers featuring specific properties. Fundamental aspects for designing polymeric architectures are, e.g., defined molecular weight, low polydispersity and high end-group fidelity. Free radical polymerization (FRP), which generally offers little control over the polymerization towards these fundamental parameters, may be adopted to obtain macromolecules with high end-group fidelity – at least related to a certain type of monomers: acrylates.

The occurrence of so-called mid-chain radicals (MCRs) in FRP of acrylates provides the basis for the facile synthesis of macromonomers, taking advantage of the follow-up reactions of the MCRs resulting in polymers with an unsaturated terminus. The obtained macromonomers represent versatile synthetic building blocks for constructing polymeric architectures.

In the present thesis the one-pot – one-step auto-initiated high-temperature polymerization method is utilized for the generation of a macromonomer library based on various acrylates and acrylate-type monomers. An extension of the macromonomer library was achieved by the synthesis of dendronized acrylates subsequently subjected to the high-temperature polymerization process.

In addition, the potential use of macromonomers as synthetic building blocks was investigated based on *n*-butyl acrylate macromonomer (BAMM) – as its purity exceeds 82 % estimated by electrospray ionization mass spectrometry (ESI-MS). Copolymeric structures – statistical copolymers with pendant side chains as well as grafted block copolymers – were synthesized via free radical copolymerization and ring-opening polymerization (ROP), respectively. The former mentioned architectures were accomplished by FRP with benzyl acrylate as the co-monomer. In addition to the detailed characterization of the obtained structures, estimated (terminal model) reactivity ratios associated with the copolymerization were determined. Grafted block copolymers were synthesized via ROP of ε-caprolactone. The employed macroinitiator was prepared by dihydroxylation of the olefinic terminus of BAMM. Possible grafting on the macroinitiators' backbone arises from transesterification side reactions during ROP.

The obtained complex polymeric materials were fully characterized via state-of-the-art polymer characterization methods, including ESI-MS hyphenated with size exclusion chromatography (SEC/ESI-MS) and liquid chromatography at critical conditions (LCCC).

Contents

1

Introduction

1.1 Motivation

Nature produces polymers since billions of years. Polysaccharides, polypeptides and polynucleotides – as representatives for natural polymer products – are considered to be the first known macromolecules considerably before the term *macromolecule* was introduced.

The early beginning of modern polymer science traces back to Henri Braconnot who synthesized nitrocellulose, the basis of celluloid, by modification of the natural polymer cellulose in the 1830s. The first example of a synthetic polymer was reported in the beginning of the 20th century by Leo Baekeland. His thermosetting phenol formaldehyde resin became famous as Bakelite.

The outstanding development of polymer synthesis was primarily driven on the back of a single scientist – Hermann Staudinger – who revolutionized the understanding of the molecular nature of polymers which was still believed to occur via the aggregation of molecules. In 1922 he proposed the concept of polymer formation by covalent bonds between single monomer molecules and thus first introduced the term *macromolecule*. Staudinger's theory laid the foundation of modern polymer science which was awarded with the Nobel Prize in 1953.

In the 1930s, polymers, synthesized from monomers according to Staudinger's theory, were made famous by Wallace Carothers. Together with his researchers at DuPont he synthesized the first synthetic rubber, nowadays known as neoprene, and a nylon polyester as a synthetic silk substitute. Additional important contributions to polymer synthesis were made by Ziegler and Natta in the 1950s by the introduction of stereospecific polymer synthesis.

To date, polymer science receives enormous interest and further enhancements regarding synthetic strategies, material properties and wide applicability in the focus. The industries' aspiration to replace classic materials of daily life by high performance polymers requires fundamental research for a selective development of new polymer-based materials.

Two basic synthetic strategies are applicable for polymer formation: step-growth polymerization and chain-growth polymerization. The latter strategy, particularly free radical polymerization (FRP) gives rise to a wide variety of polymeric materials according to monomer type, achievable molecular weights and polydispersities. However, conventional FRP enables very little control of the polymerization process over molecular weight, polydispersity and macromolecular architecture based on highly reactive monomer radicals. Improvement of control in FRP was obtained by the development of so-called controlled radical polymerization methods (CRP). Nowadays widely accepted CRP methods include nitroxide-mediated polymerization (NMP),[1] atom transfer radical polymerization (ATRP)[2] and reversible addition fragmentation chain transfer (RAFT) polymerization.[3-4]

FRP processes can be utilized to generate defined polymer termini. Several monomers, especially acrylates, deviate from ideal FRP which facilitates the synthesis of functionalized polymers. The main focus of the present thesis is placed on the synthesis of macromonomers through taking advantage of the occurrence of so-called mid-chain radicals (MCR) in high-temperature free radical polymerization.[5-8]

During the free radical polymerization of acrylates, inter- and intramolecular transfer to polymer generates tertiary propagating radicals – MCRs – from secondary propagating radicals (SPR).[9] As a consequence of the occurring MCR, branching in poly(acrylate)s can be observed, such as long chain branching from intermolecular transfer and short-chain branching from the so-called backbiting. The transfer reactions occurring during the polymerization can be utilized for macromonomer formation. Macromonomers feature an unsaturated endgroup which can potentially act as a reaction point in further polymerizations.

The significant follow-up reaction of the MCR for the formation of macromonomers is the so-called β-scission resulting in an unsaturated endgroup. The synthesis can be directed towards the macromonomer formation through the application of high temperatures (above 100°C) and low monomer concentration; even at ambient temperature already 60% MCR exist.[5] Macromonomer synthesis via the high-temperature polymerization method was first introduced by Chiefari *et al.*[10]

Macromonomers with their unsaturated character can subsequently be utilized as versatile synthetic building blocks to provide access to well-defined polymeric architectures such as branched, block-, graft-, comb- or star-polymers.

1.2 Thesis Overview

The present thesis addresses the synthesis of macromonomers via the high-temperature polymerization method and their behavior in further reaction sequences to access well-defined polymeric architectures. A brief outline of the conducted synthesis in this work is illustrated in Figure 1-1.

Based upon *n*-butyl acrylate macromonomer (BAMM), as the initially synthesized macromonomer via this approach, an extended macromonomer library has been prepared. Various macromonomers were synthesized via the high-temperature acrylate polymerization route based on methyl (MA), ethyl (EA), *n*-butyl (BA), *t*-butyl (*t*BA), 2-ethylhexyl (2-EHA), isobornyl (iBoA) and 2-[[(butylamino)carbonyl]oxy]ethyl (BECA) acrylate. Additionally, the self-initiated macromonomer formation has been further investigated by an in-depth kinetic evaluation.[11] A further extension of the macromonomer library was achieved by the introduction of dendritic structures. Dendronized acrylates based on 2,2-bis(hydroxymethyl)propionic monomer (bis-MPA) were synthesized through modular ligation of convergently synthesized acetylene bearing dendrons with azido-acrylates. Subsequent macromonomer formation for all prepared generations was investigated.[12]

The potential use of macromonomers in further polymerizations was investigated by the formation of copolymeric structures. BAMM was copolymerized with benzyl acrylate as co-monomer in a free radical copolymerization. A (terminal) Mayo-Lewis analysis was conducted based on the experimental copolymer compositions resulting in estimated reactivity ratios for the underlying copolymer system.[13]

Additionally, BAMM could be employed as a ring-opening polymerization (ROP) initiator after successful transformation of the unsaturated terminus into a macroinitiating moiety via a dihydroxylation process. The initiation efficiency of the macroinitiator in ROP with organo, metal and enzyme catalysis was investigated.[14]

A detailed image of the generated copolymers poly(*n*-butyl acrylate)-*b*-poly(benzyl acrylate) and poly(*n*-butyl acrylate)-*b*-poly(ε-caprolactone) was achieved by the application of various characterization methods. Predominantly, electrospray ionization mass spectrometry (ESI-MS) and ESI-MS hyphenated with size exclusion chromatography (SEC/ESI-MS) was applied to the synthesized structures, as well as liquid chromatography at critical conditions (LCCC). Conventional SEC and nuclear magnetic resonance (NMR) spectroscopy were further utilized to determine the structure of the copolymers.

In summary, the current thesis provides insights into high-temperature macromonomer synthesis and the use of the resulting oligomeric building blocks for constructing polymer architectures.

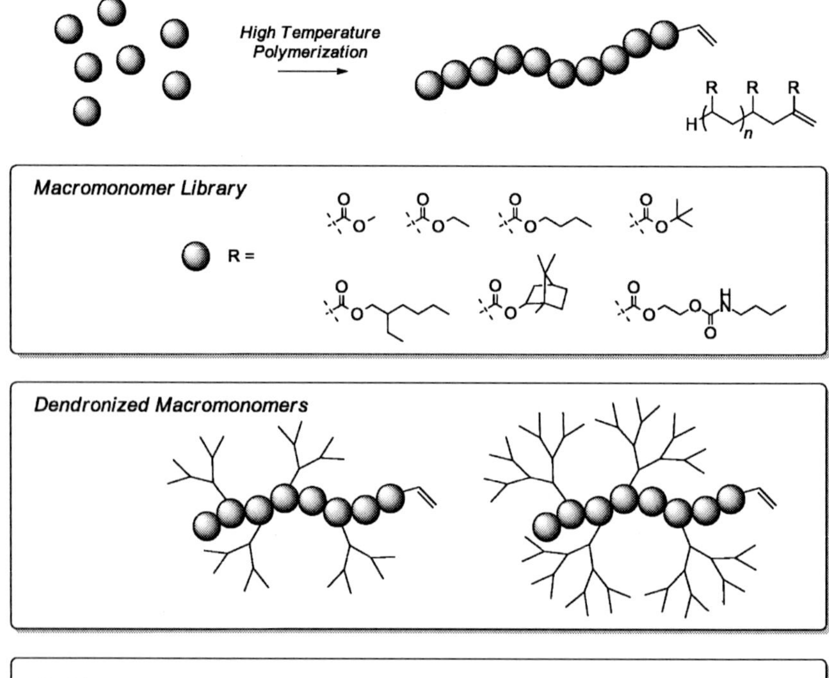

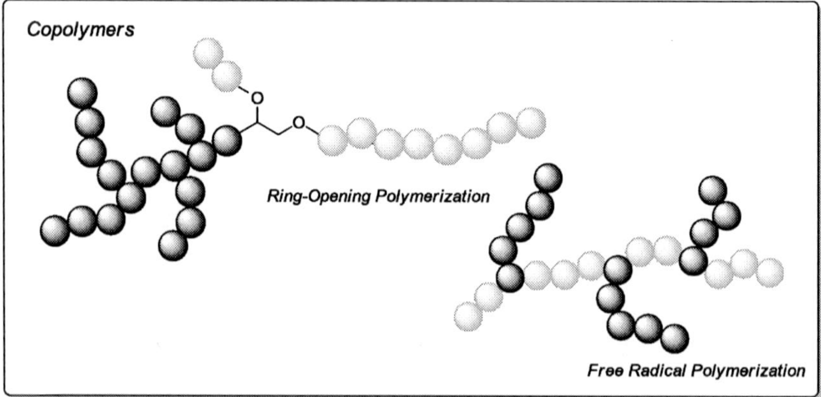

Figure 1-1. Outline of the synthesis of various polymeric architectures from macromonomeric building blocks.

2

Polymerization Techniques

2.1 Free Radical Polymerization

Conventional free radical polymerization (FRP) allows for the formation of high molecular weight polymers by a chain reaction process. Any (activated) monomer with an unsaturated functionality can be utilized for a free radical polymerization. In general, three characteristic steps occur during chain polymerization: initiation, propagation and termination.

Reactive radicals for polymerization initiation are generated from an initiator I. The decomposition of the initiator generates two active radicals R· (either symmetrical or unsymmetrical) where the decomposition rate coefficient of the first order reaction is given by k_d.

$$\mathrm{I} \xrightarrow{k_d} 2\,\mathrm{R}\cdot \qquad (2\text{-}1)$$

For initiation several compounds can be used, e.g., peroxo- or azo-initiators. The decomposition of the initiator can either be thermally or photolytically triggered. The driving force of the decomposition is often due to the elimination of gases (entropically favored). The *in-situ* generated radicals further attack the electron rich π-system of the vinyl group in the monomer M to transfer the radical center to the monomer.

$$R_i\cdot + M \xrightarrow{k_p^i} R_{i+1}\cdot \tag{2-2}$$

The propagation step is repeated several times resulting in a macro-radical species $R_{i+1}\cdot$ with the radical center on the terminus. The propagation rate coefficient k_p describes the propagation for every step apart from potential chain length dependence.

The polymerization can be stopped through termination reactions. Two types of termination pathways need to be distinguished. In both cases two radical centers react with each other to form a single bond. Equation (2-3) depicts the termination reaction via recombination, where two radicals with different chain length i and j recombine to form an inactive or dead polymer chain P_{i+j}. The chain length of the obtained dead polymer equals the sum of both participating macro-radicals. Aside from recombination, disproportionation can occur as shown in equation (2-4). In disproportionation the chain length of the dead polymers is equal to the chain length of the macro-radicals. The termination reaction occurs via a hydrogen abstraction and thus one polymer chain is equipped with an unsaturated terminus.

$$R_i\cdot + R_j\cdot \xrightarrow{k_t^{i,j}} P_{i+j} \tag{2-3}$$

$$R_i\cdot + R_j\cdot \xrightarrow{k_t^{i,j}} P_i + P_j \tag{2-4}$$

Termination reactions in a free radical polymerization are known to be diffusion controlled.[15] The rate of termination reduces with increasing conversion and consumption of monomer due to higher viscosity and chain entanglement. The termination process is associated with an isothermal auto acceleration, the so-called gel-effect or Norrish-Trommsdorff effect, at high conversion.[16-17]

Additional to the above mentioned reaction steps, potential side reactions of the reactive radicals in FRP have to be considered. In FRP chain transfer reactions are observed which transfer the radical character from a propagating macro-radical to another species present in the polymer mixture. The propagating radical is terminated by a transfer agent X with a transfer rate coefficient k_{tr} depicted by equation (2-5). Simultaneously a new radical is generated which subsequently reinitates the polymerization (2-6).

$$P_i\cdot + X \xrightarrow{k_{tr}} P_i + X\cdot \tag{2-5}$$

$$X\cdot + M\cdot \xrightarrow{k_{rei}} P_n \tag{2-6}$$

Chain transfer to monomer can occur from the monomer itself, polymer chains generated during the polymerization process, solvent and deliberately added transfer agents. Transfer agents can thus selectively employed to reduce the molecular weight of the polymer or to introduce defined endgroups.

Transfer reactions, however, limit the maximum molecular weight that may be achieved in FRP. The degree of polymerization is decreased while the amount of polymer chains is increased. The effectiveness of chain transfer processes is quantified by the chain transfer constant C_X – the ratio of the transfer rate coefficient to the propagation rate coefficient. Chain transfer to polymers, however, involves the abstraction of a hydrogen atom from the polymer backbone. Branching of the polymers is the consequence of this reaction step. A detailed discussion of chain transfer to polymers can be found in Chapter 3.

The free radical polymerization is applicable to a variety of monomers, however, a substantial drawback of the mentioned polymerization technique is the limited control over molecular weight, polydispersity, endgroup fidelity and architecture of the polymers. In the following, developments and improvements related to radical based polymer synthesis are shown in detail.

2.2 Living/Controlled Free Radical Polymerization

In 1956, Szwarc first introduced the concept of living polymerization.[18] He showed that anionic polymerization technique features molecular chain growth without termination steps resulting in efficient control over architecture and structure of the polymers. The pioneering work of Szwarc was subsequently followed by continuous enhancement of controlled polymerization till this day.

Living/Controlled *radical* polymerization (CRP) protocols are based on an equilibrium between dormant and active species. During the exchange between the two species, chain propagation can occur with limited termination reactions. The most common CRP protocols are nitroxide-mediated polymerization (NMP),[1] atom transfer radical polymerization (ATRP)[2] and reversible addition fragmentation chain transfer (RAFT) polymerization.[3-4] The control of polymerization is based on two basic mechanisms. NMP and ATRP are controlled through a reversible termination step, whereas control in RAFT polymerization profits from degenerative chain transfer. The above mentioned polymerization protocols are outlined below.

2.2.1 Nitroxide-Mediated Polymerization (NMP)

In NMP stable nitroxide radicals are used for controlling the polymerization. NMP represents a specific type of stable free radical polymerization (SFRP).[19] The mediating nitroxide radicals reversibly terminate the propagating chains via the formation of an alkoxyamine, thus suppressing the bimolecular termination reaction. This so-called persistent radical effect (PRE) forms the basis of NMP.[20] The persistent radicals terminate by reversible radical-radical coupling and the chain propagation continues after reversing the termination step.

The propagation occurs only through the reactive radical species by radical coupling to monomer species. The general mechanism of NMP is shown in Scheme 2-1.

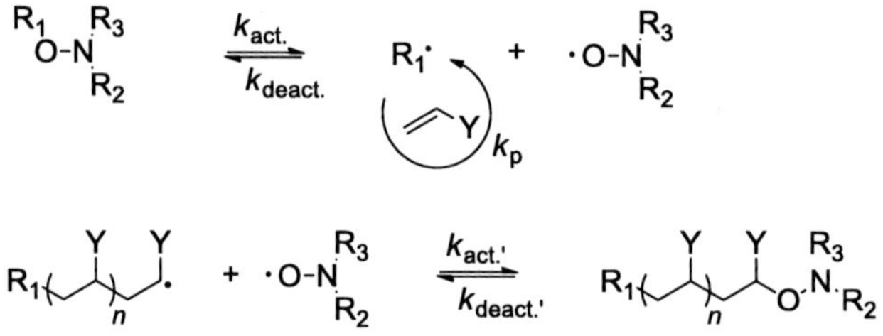

Scheme 2-1. General mechanism of nitroxide-mediated polymerization (NMP).

The persistent nitroxide radical can be introduced into the polymerization by two pathways. On the one hand, a stable nitroxide radical such as 2,2,6,6-tetramethylpiperidine-N-oxyl (TEMPO) can be applied together with a conventional initiator which initiates the polymerization. On the other hand, an alkoxyamine which thermally decomposes to a radical fragment and the stable nitroxide is used. Both methods lead to reactive radicals initiating the polymerization and a stable nitroxide which reversibly terminates the propagating chains.

2.2.2 Atom Transfer Radical Polymerization (ATRP)

Advantages of ATRP compared to NMP method are the compatibility towards a variety of monomers and lower polymerization temperatures. Equally to NMP, control in ATRP is achieved through a reversible termination step. The radical formation occurs through a transition metal catalyst (commonly CuBr) undergoing a reversible redox process. Reactive radicals stem from halogen abstraction from an organic halide – the initiator – to the transition metal. The addition of a ligand (L) increases the solubility of the transition metal salt in the organic medium. The reactive radicals referred to as 'activator' initiate the polymerization. Contrary, the transition metal salt in higher oxidation state (Cu^{II}) reduces the concentration of propagating radicals by deactivation and thus minimizes the bimolecular termination reaction. The oxidized transition metal behaves as stable radical and acts as 'deactivator'. The general mechanism of ATRP is presented in Scheme 2-2. The equilibrium between propagating active species and dormant species must be adjusted thus that the activation rate coefficient is sufficiently large to suppress the bimolecular termination of propagating radicals, yet allowing propagation of active polymer chains.

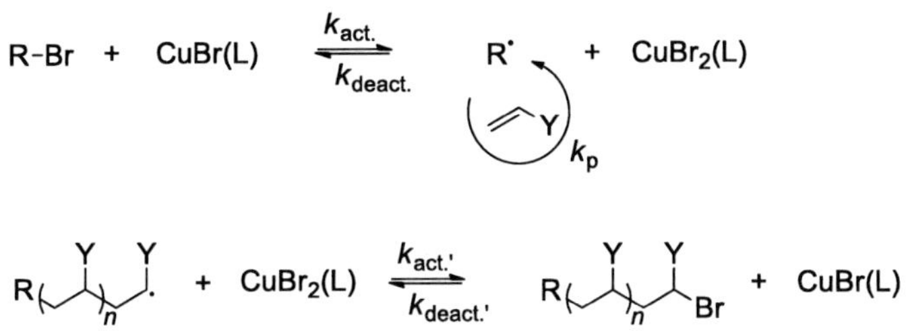

Scheme 2-2. General mechanism of atom transfer radical polymerization (ATRP).

2.2.3 Reversible Addition Fragmentation Chain Transfer (RAFT)

In contrast to NMP and ATRP, the control of chain growth in RAFT is based on reversible chain transfer processes. A chain transfer agent – RAFT agent – is introduced into the polymerization which transfers a group to a propagating radical forming a dormant species. Common RAFT agents are dithioesters with variable substituents Z and R. The choice of Z and R is based on the chemical nature of the monomer and determines the transfer constants. The Z group activates or deactivates the thiocarbonyl double bond of the RAFT agent and thus modifies the stability of the generated intermediate radicals. The R group acts as a leaving group and determines the stability of the intermediate radicals (pre-equilibrium) and reinitiates macromolecular growth.[21]

Conventional initiators are applied in RAFT polymerization for generating primary radicals. In the pre-equilibrium the propagating radical and the RAFT agent are converted into an intermediate radical followed by fragmentation to a macromolecular RAFT agent and a radical species. Further propagation is then achieved by a reinitiation process followed by the main equilibrium. The control of the polymerization is reached through dormant and activated species in equilibrium and minimizing the occurrence of bimolecular termination events for the individual radical species. The basic mechanistic aspects are presented in Scheme 2-3.

In NMP and ATRP a low propagating radical concentration suppresses the termination reaction thus controlling the polymerization. In RAFT polymerization the overall radical concentration, however, is not lowered and the rate of termination is close to identical to a conventional FRP process with the same initiator concentration. Once a propagating radical is set to the dormant state, another propagating radical is released.

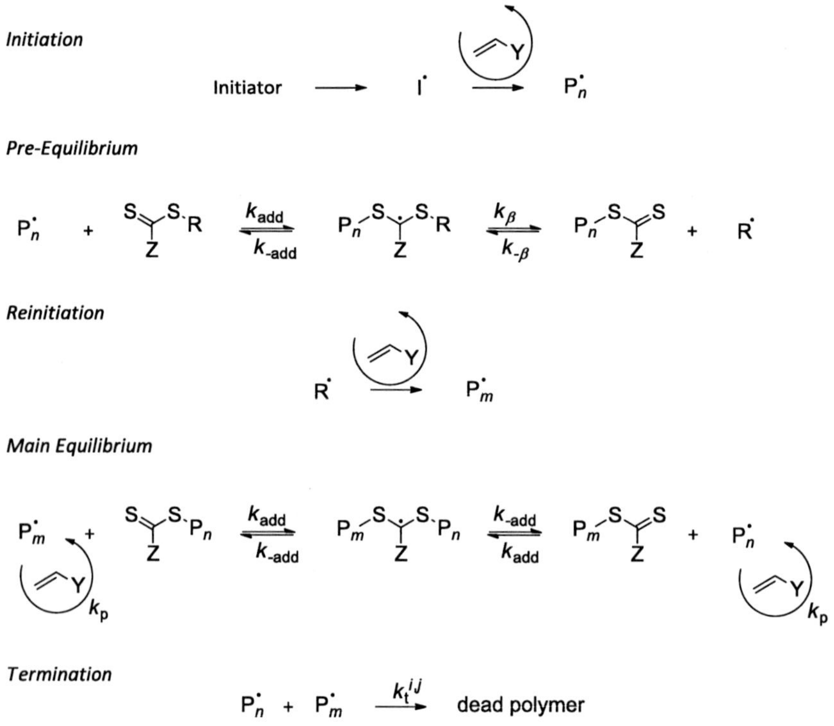

Initiation

Initiator ⟶ $I^{\bullet}$ ⟶ $P_n^{\bullet}$

Pre-Equilibrium

$P_n^{\bullet}$ + S=C(Z)–S–R $\underset{k_{-add}}{\overset{k_{add}}{\rightleftharpoons}}$ P_n–S–C$^{\bullet}$(Z)–S–R $\underset{k_{-\beta}}{\overset{k_{\beta}}{\rightleftharpoons}}$ P_n–S–C(Z)=S + $R^{\bullet}$

Reinitiation

$R^{\bullet}$ ⟶ $P_m^{\bullet}$

Main Equilibrium

$P_m^{\bullet}$ (k_p) + S=C(Z)–S–P_n $\underset{k_{-add}}{\overset{k_{add}}{\rightleftharpoons}}$ P_m–S–C$^{\bullet}$(Z)–S–P_n $\underset{k_{add}}{\overset{k_{-add}}{\rightleftharpoons}}$ P_m–S–C(Z)=S + $P_n^{\bullet}$ (k_p)

Termination

$P_n^{\bullet}$ + $P_m^{\bullet}$ $\xrightarrow{k_t^{ij}}$ dead polymer

Scheme 2-3. General mechanism of the reversible addition fragmentation chain transfer (RAFT) process.

3

Macromonomers – Theory and Concept

3.1 Chemical Structures of Macromonomers

Macromonomers are oligomeric or polymeric species with a polymerizable terminus.[22-26] Thus, macromonomers can be utilized as versatile synthetic building blocks to provide access to well-defined polymeric architectures such as branched, block-, graft-, comb- or star-polymers.[27-28] Several synthetic pathways are established for the achievement of ω-functionalized macromonomers.[29]

The chemical structure of a terminal double bond – the basic characteristic of macromonomers – varies depending on the synthetic strategy. As shown in Figure 3-1, macromonomer **1** features a 1,1'-disubstitution on the terminus, which is referred to as a geminal double bond. Additionally, polymeric structure **2** with a 1,2-disubstituted terminus (vicinal double bond) can be achieved, and less common – yet conceivable – is a vinyl group **3** which features a monosubstitution on the olefin terminus.

As already stated, the following work will focus on macromonomers of structure **1**, which have been prepared via high-temperature acrylate polymerization in order to be further utilized for constructing various polymeric architectures.

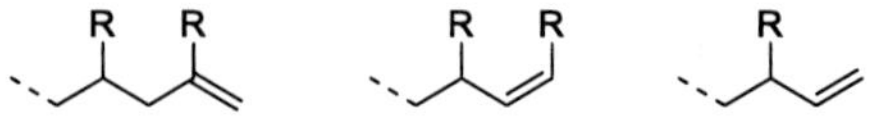

Figure 3-1. Basic types of macromonomers. The character of the vinyl terminus can be of disubstituted nature (geminal **1** and vicinal **2**) or monosubstituted nature **3**.

3.2 The Origin of Macromonomers

3.2.1 Free Radical Polymerization of Acrylates

The kinetics of conventional free radical polymerization outlined in Chapter 2 are described through a first order rate law in terms of monomer concentration (3-1).

$$R_{\mathrm{P}}=\frac{\mathrm{d}[\mathrm{M}]}{dt}=-\frac{k_{\mathrm{p}}}{\sqrt{k_{\mathrm{t}}}}\cdot[\mathrm{M}]\cdot\sqrt{2k_{\mathrm{d}}\cdot f\cdot[\mathrm{I}]} \tag{3-1}$$

The rate of polymerization R_{P} is described by the initiation efficiency f, initial monomer concentration [M], initiator concentration [I] and the rate coefficient of decomposition k_{d} as well as the propagation k_{p} and termination k_{t} rate coefficients. Inspection of the rate law for conventional free radical polymerization suggests the following concentration dependence:

$$R_{\mathrm{p}} \propto [\mathrm{M}]^{\omega}\cdot[\mathrm{I}]^{0.5} \tag{3-2}$$

For conventional free radical polymerization ω equals unity, however, for the polymerization of acrylates this value deviates significantly.

In the literature, a reaction order greater than unity in terms of monomer concentration has been observed by several research groups.[30-40] For a deeper understanding of the kinetics in free radical polymerization, the propagation rate coefficient k_{p} has been widely studied for an elucidation of the detailed mechanistic underpinning in acrylate polymerizations. The determination of k_{p} can be achieved via the pulsed laser polymerization (PLP) technique and subsequent analysis via size exclusion chromatography (SEC). The PLP-SEC technique has been recommended by the IUPAC Working Party on *Modeling of Polymerization Kinetics and Processes* after the pioneering work of Olaj in 1987 within this research area.[41-42] For a detailed description of the PLP-SEC technique refer to the review of Beuermann and Buback.[43] This technique for the propagation rate coefficient determination has been applied for several monomers such as (functional) styrenes and (meth)acrylates. A selection of k_{p} values of styrene, methyl methacrylate and *n*-butyl acrylate at 20°C are given in Table 3-1.

Table 3-1. Secondary propagating radical k_p values for selected monomers determined via PLP-SEC measurements.

monomer	A / L·mol^{-1} s^{-1}	E_A / kJ·mol^{-1}	k_p (20 °C) / L·mol^{-1}·s^{-1}
styrene[44]	4.27×10^7	32.5	69
methyl methacrylate[45]	2.67×10^6	22.4	272
n-butyl acrylate[46]	2.24×10^7	17.9	14476

The monomers in Table 3-1 illustrate that the propagation rate coefficient k_p of *n*-butyl acrylate – and indeed all other acrylates – is significantly higher compared to styrene or methyl methacrylate at ambient temperature. However, the measurement of acrylates was found to be challenging at temperatures above 20 °C.[47-51] The addition of monomer to the propagating macro-radical is observed to be fast, however, propagation rate coefficients of acrylates were found to be lower than actually expected.[46-47,52-53] Chain transfer to monomer, low termination to propagation ratio and intramolecular chain transfer to polymer were postulated to be the reasons for these difficulties.[48,54] Molecular weight distributions resulting from 100 Hz frequency (and temperatures above 20 °C) for acrylates show either no or a broadened SEC-PLP profile. At high temperatures, however, k_p describes the propagation rate coefficient of the secondary propagating radicals and not the effective k_p for the polymerization. Recently, Barner-Kowollik *et al.* further investigated SEC-PLP with a high frequency pulsing (500 Hz) allowing the determination of k_p up to 70 °C.[55] The determination of k_p was further investigated by Willemse *et al.* via PLP coupled to matrix assisted laser desorption ionization (MALDI) time of flight (ToF) mass spectrometry.[56] Several reasons for the observed deviations in PLP-SEC experiments of acrylates have been hypothesized, such as transfer to monomer, polymer and solvent.[36,40,50,57-58]

Initially, the work of Scott and Senogles brought elucidation to the unexpected behavior of acrylates during a free radical polymerization.[33-35] The hypothesis was that intramolecular transfer to polymer is a potential cause for the unexpected behavior. The appearance of mid-chain radicals (MCRs) from secondary propagating radicals (SPR) (for chemical structures see Figure 3-2) was evidenced by the use of electron spin resonance (ESR) spectroscopy.[5-8]

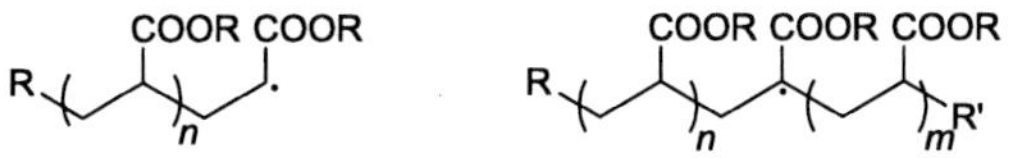

COOR COOR
R n
(SPR)

COOR COOR COOR
R n m R'
(MCR)

Figure 3-2. Chemical structures of secondary propagating radicals and mid-chain radicals from intramolecular transfer to polymer.

Furthermore, Ahmad *et al.* observed quaternary carbons in poly(*n*-butyl acrylate) via ^{13}C nuclear magnetic resonance (NMR) spectroscopy which arise from the propagation of a tertiary MCR forming chain branches.[59] The formation of chain branches in the polymer backbone was found to be dominant in diluted systems and thus intramolecular transfer is more favored in solution ($[M]_0 \leq 10$ (w/w)). The level of branching and quantity of MCRs is not only a function of temperature but also a function of monomer concentration.[59] The intramolecular transfer reaction was already well known from high-temperature ethylene polymerization as well as the copolymerization of ethylene and *n*-butyl acrylate.[58,60] It was also evidenced that intramolecular transfer is not occurring during methacrylate polymerization and thus represents a specific feature of acrylates.[55]

Based on the initial pioneering work of Scott and Senogles the free radical polymerization of acrylates was clarified during the past decades. Kinetic equations based on the present knowledge have been published.[61] In the following section the occurrence of the mid-chain radicals and its further reaction pathways are discussed.

3.2.2 Insights into Mid-chain Radicals and Backbiting

3.2.2.1 Intermolecular Transfer to Polymer

Above all, the intermolecular transfer between two independent polymer chains is regarded to be the simplest type of transfer.[9] During intermolecular transfer, the hydrogen atom abstraction occurs by a secondary propagating radical from another polymer chain. The chemical driving force of this transfer reaction is radical stability, as a tertiary radical is more stable than a secondary radical. Scheme 3-1 depicts intermolecular transfer which results in MCRs. The growing chain with the radical chain end is terminated during the transfer process, while conversely a previously dormant polymer chain is reactivated, continuing the polymerization at a random position of the polymer backbone. The obtained MCR carries both endgroups of the previous dormant polymer chain. The transfer can occur randomly at any position to a quaternary carbon equipped with an ester moiety within the backbone.

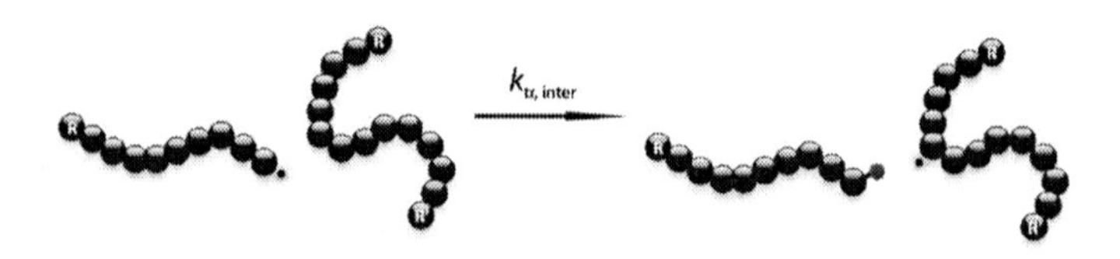

Scheme 3-1. Intermolecular transfer of a secondary propagating radical forming a mid-chain radical.[9]

The degree of branching depends on the rate of intermolecular transfer which again is highly affected by the monomer concentration during polymerization correlated with conversion. The rate law of the intermolecular transfer is given by equation (3-3),

$$v_{\text{tr,inter}} = k_{\text{tr,inter}} \cdot c_{\text{R}} \cdot c_{\text{U}} \tag{3-3}$$

where $k_{\text{tr,inter}}$ denotes the rate coefficient of the intermolecular transfer reaction and c_{R} and c_{U} are the propagating radical concentration and the quantity of available backbone repeat units, respectively. Hence, intermolecular transfer is a bimolecular reaction. The assumption of non-diffusion control of the transfer reaction is imbedded within the equation above.

3.2.2.2 Random Intramolecular Transfer

In addition to an intermolecular transfer process, the propagating chain radical is capable of undergoing intermolecular transfer within the same polymer chain.[9] The schematic of the intramolecular transfer is shown in Scheme 3-2.

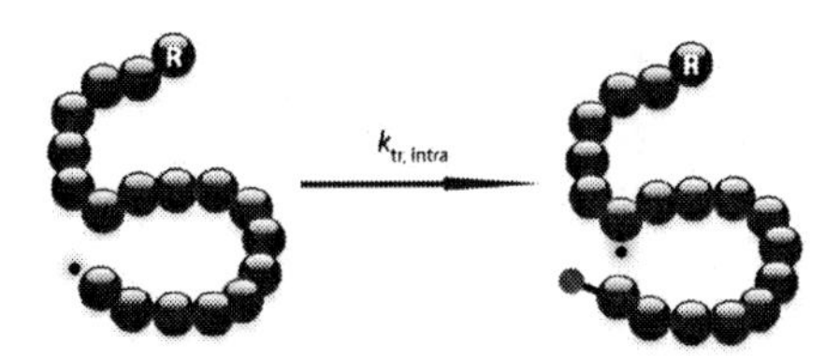

Scheme 3-2. Intramolecular transfer from a secondary propagating radical to form a mid-chain radical by hydrogen atom abstraction from a random position of the polymer backbone.[9]

The position of hydrogen abstraction on the polymer backbone is random and changes the radical from a SPR to a MCR. The MCR is more stable due to a tertiary radical center therefore giving rise to branching during polymerization. In comparison to intermolecular transfer, the generated MCR is equipped with only one specific endgroup. The second endgroup – a hydrogen – is obtained from hydrogen atom abstraction. The reaction rate of the random intramolecular transfer reaction is given by equation (3-4).

$$v_{\text{tr,intra}} = k_{\text{tr,intra}} \cdot c_{\text{R}} \tag{3-4}$$

Here, $k_{\text{tr,intra}}$ describes the transfer rate coefficient of the unimolecular reaction. The hydrogen atom abstraction can occur at any position of the polymer backbone. A restriction only needs to be added for short chains during the polymerization. A certain length of the

SPR needs to be reached before the intramolcular transfer can occur. During coil formation – as the chain propagates – the SPR is more likely to be located within the coil and the intramolecular transfer reaction is therefore favored due to the proximity of the radical center to the polymer backbone.

However, during intramolecular transfer some positions on the polymer backbone are favored in comparison to others, as discussed below.

3.2.2.3 Intramolecular Transfer via a 6-Membered Ring (Backbiting)

As discussed above, the intramolecular transfer is favored with regard to specific positions on the polymer backbone.[9] One favorable process is intramolecular transfer forming a six-membered ring transition structure associated with a 1,5-hydrogen shift. The schematic is depicted in Scheme 3-3.

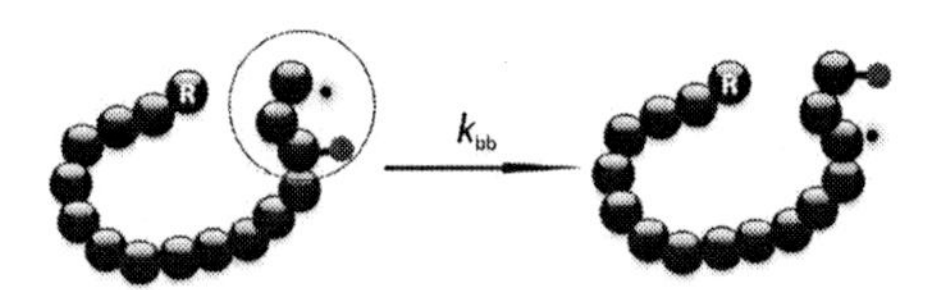

Scheme 3-3. Intramolecular transfer from a secondary propagating radical to form a mid-chain radical favored by a 6-membered ring (backbiting).[9]

The radical center of the resulting MCR is located at the 5th carbon (third repeat unit) of the backbone counted from the SPR radical end. The chemical structures of the transfer are highlighted in Scheme 3-4. The former hydrogen atom is transferred to the last repeat unit and becomes the endgroup.

ROOC COOR H COOR → ROOC COOR H COOR

Scheme 3-4. Formation of MCR via a six-membered ring structure.[9]

The specific transfer via a six-membered ring structure is termed backbiting and, thus, should be distinguished from other types of intramolecular transfer reactions. Backbiting is an unimolecular reaction according to the random intramolecular transfer albeit with a backbiting rate coefficient $k_{tr,bb}$. The reaction rate is given by equation (3-5).

$$v_{tr,bb} = k_{tr,bb} \cdot c_R \tag{3-5}$$

Intramolecular transfer is not only restricted to the six-membered transition state, however backbiting is favored due to the highest energetic benefit. Generally, various transition states are conceivable for the intramolecular transfer. The difference in backbiting and transfer via higher (6+2*n*)-membered ring structures is expressed by the type of branching. While backbiting leads to short chain branching, the other transfer reactions lead to long chain branching. In addition to the noted transfer reactions, Kajiwara postulated further transfer of the MCRs through a 1,5-hydrogen shift.[62-63]

3.2.3 Potential Reactions of MCR and the Formation of Macromonomers

Willemse *et al.* determined the quantity of MCRs occurring during PLP of *n*-butyl acrylate via ESR spectroscopy. They observed that even at ambient temperature 60 % MCR exist.[5] The experiment confirmed the expectations since tertiary MCR are more stable and of a less reactive nature compared to secondary propagating radicals.

Several reactions of MCRs during polymerization are conceivable: (a) propagation, (b) termination and (c) β-elimination/scission. Reaction pathways (a) and (b) are analogous to the secondary propagating radicals. During propagation monomer units are added to the propagating chain end. Unlike SPRs, MCRs induce branching during propagation. For termination both recombination and disproportionation occur. The most relevant follow-up reaction for the macromonomer formation, however, is the so-called β-elimination/scission, or shorter β-scission. The occurrence of MCR and thus β-scission was evidenced by Yamada and co-workers.[6-7,10,49] Depending on the type of MCRs endgroups (referred to either as inter- or intramolecular transfer reaction) two species are formed from β-scission. The proposed mechanism of the β-scission is illustrated in Scheme 3-5.[22,64-65]

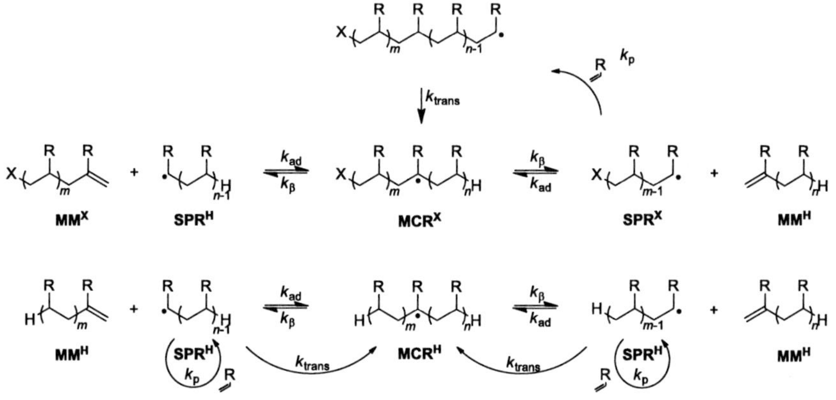

Scheme 3-5. Mechanistic details of the formation of macromonomers from mid-chain radicals undergoing β-scission.

Interestingly, the reaction mechanism does not only entail simple consecutive reactions, (i.e., the formation of the MCR and the scission reaction) but consists of a complex set of reaction equilibria by which MCRs are also formed via the addition of propagating radicals to macromonomer, a reaction that becomes progressively more prominent with increasing monomer to macromonomer conversion.[64] In the top row of Scheme 3-5, the MCR is initiated by endgroup X, whereas the second row reflects a hydrogen atom terminated MCR. The macromonomer, as one species, is obtained with an unsaturated endgroup and the initiation fragment X (**MM^X** vs. **MM^H**). Additionally, a secondary propagation radical results from β-scission (**SPR^H** vs. **SPR^X**). Both species are in equilibrium with the MCR and are therefore participants of a reversible reaction. The formation of MCRs via backbiting and the subsequent β-scission to form macromomers shows a high activation energy and thus demands the use of elevated temperatures for the macromonomer formation process. With increasing temperature and low monomer content the majority of secondary propagation radicals are transferred into MCRs.[5]

The reaction mechanism has been well-studied via a detailed mass-spectrometry analysis[66-69] and detailed kinetic simulations, which were carried out to elucidate the limitations of macromonomer production in this approach.[64] Such a direct route enables the synthesis of highly pure macromonomers via a facile one-pot – one-step procedure.[11]

3.3 High-Temperature Acrylate Polymerization

While living/controlled polymerization techniques are commonly used to generate polymers with homogeneous endgroups, other methods to yield macromonomers employing conventional radical polymerizations exist by which similar levels of endgroup fidelity can be achieved. In the late 1990s Chiefari *et al.* introduced a method to create acrylic macromonomers in high yields by polymerizing at high temperatures, albeit with higher polydispersity as compared to living/controlled radical polymerization methods.[10,70] The general procedure to obtain macromonomers from high-temperature acrylate polymerization is represented in Scheme 3-6.

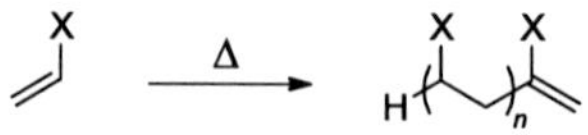

X = COOR

Scheme 3-6. General reaction pathway for high-temperature acrylate polymerization.

Styrenic and acrylic macromonomers were achieved in high yields employing an appropriate solvent, initiator and monomer concentration at temperatures ranging from 80 °C to 240 °C.

The structure of the macromonomers was confirmed via NMR spectroscopy. The mechanism makes use of the transfer reactions which take place during acrylate polymerizations where so-called mid-chain radicals (MCR) are formed.[5,9,61] Those MCRs subsequently undergo scission reactions forming macromonomeric species, if the reaction temperature is sufficiently high (T > 120 °C). At lower temperatures, monomer addition is favored over scission, thus forming chain branches rather than a defined unsaturated endgroup.

The reaction mechanism of high-temperature macromonomer formation has been elucidated by Barner-Kowollik and co-workers via a detailed mass spectrometric analysis.[68,71] A few years later Junkers *et al.* refined the synthetic procedure, obtaining highly pure macromonomers in a one-pot – one-step procedure from a monomer/solvent mixture without addition of any controlling agent or initiator.[67,72] Compared to other existing methods for macromonomer formation, such a direct synthetic approach is very appealing as it does not require the use of specific control agents, post-processing of polymers or purification. During the high-temperature polymerization of acrylates several chemical structures are obtained. The most important chemical structures are shown in Figure 3-3.

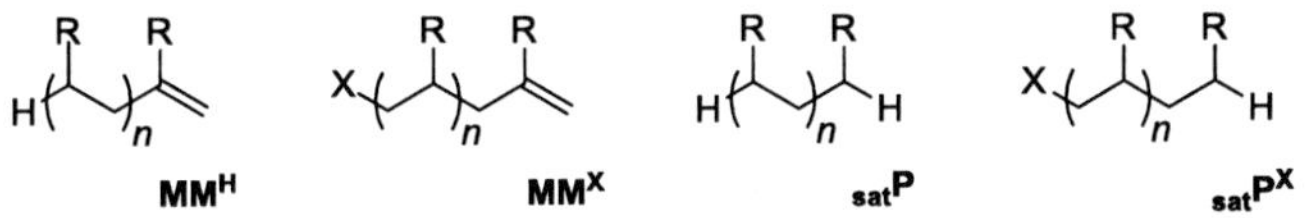

$\mathbf{MM^H}$ $\mathbf{MM^X}$ $\mathbf{_{sat}P}$ $\mathbf{_{sat}P^X}$

Figure 3-3. Chemical structures of species found during the polymerization process.

Due to the constant regeneration of propagating radicals by the β-scission reaction, macromonomer of the type $\mathbf{MM^H}$ is the most significant and abundant reaction product, even under conditions with a steady supply of initiating radicals. Macromonomers of the type $\mathbf{MM^X}$ (where X stands for any primary initiating radical fragment, such as an initiator-derived radical, transfer-to-solvent or other small radicals) may also be generated as side products. It should be noted that from a synthetic viewpoint, $\mathbf{MM^X}$ and $\mathbf{MM^H}$ are both functional products which are suitable for follow-up chemistry as both carry the same olefinic terminus. In addition to the unsaturated species, structures of the type $\mathbf{_{sat}P}$ and $\mathbf{_{sat}P^X}$ are potential species of the reaction (X is referred to be the same radical fragment as in structure $\mathbf{MM^X}$). The hydrogen atom terminated species, or saturated endgroups, cannot act as macromonomers for further chemistry. Fortunately, these dead oligomeric or polymeric species are formed in minor quantities.

The major advantage of macromonomer synthesis via high-temperature polymerization is its simplicity. The pure macromonomer is obtained in high yields in a one-pot – one-step procedure without any further purification.

3.4 Further Macromonomer Synthesis Strategies

3.4.1 Catalytic Chain Transfer (CCT) Polymerization

In the early 1970s Smirnov and Marchenko introduced catalytic chain transfer (CCT) using cobalt porphyrins as a catalyst.[73-74] In the polymerization of methyl methacrylate (MMA) control over molecular weight and unsaturated endgroups was obtained. The unsaturated terminus of the polymers was evidenced to be of olefinic nature through NMR spectroscopy,[75-76] whereby significant signals in the methacrylic region were observed. The polymers were achieved by the addition of a Co(II) low spin complexes acting as a chain transfer catalyst.[74,77-79] However, due to high costs, color and partial insolubility in organic media of the porphyrines and related derivates it became necessary to further develop the catalytic system. O'Driscoll and co-workers later synthesized cobaloxime and used it in the successful CCT polymerization of MMA.[80] The catalyst based on dimethyl glyoxime featured a higher reactivity towards CCT. Several catalysts for CCT polymerization are shown in Figure 3-4.

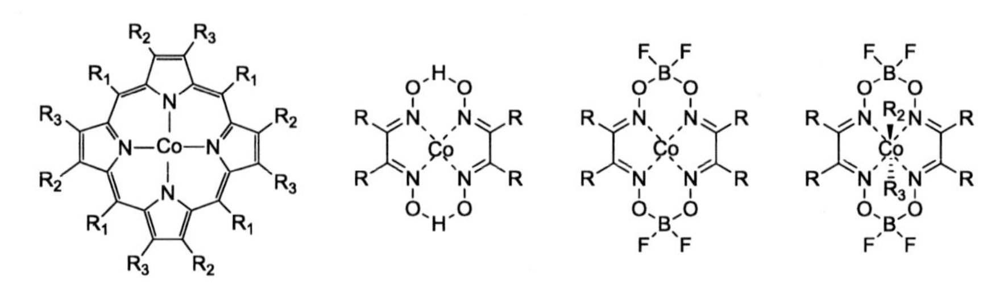

Figure 3-4. Selection of CCT catalysts: porphyrin based Co(II) catalyst, dimethyl glyoxime Co(II), dimethyl glyoxime Co(II) with BF_2-bridge and alkylated Co(III) complex.

However, Co CCT catalysts are sensitive towards oxidation and hydrolysis, which needed to be further reduced. The desired improvement was obtained by the insertion of BF_2 bridges. The sensitivity was then further reduced by Moad and colleagues by the use of alkylated Co(III) complexes, whereby the active Co(II) catalyst is generated *in-situ* by homolytic dissociation or radical induced reduction.[57] Today, the Co(II) cobaloxime complex, named COBF, is the most often employed catalyst for CCT polymerizations.

Macromonomers via CCT polymerization can be synthesized from monomers equipped with an α-methyl group such as methyl methacrylate, α-methyl styrene and methacrylnitrile. In the absence of the α-methyl group (i.e., styrene and acrylates) Co-C bond formation between the catalyst and the monomer is favored, which lowers the chain transfer constant.[81-84] The mechanism of CCT polymerization, as shown in Scheme 3-7, has been well studied by several research groups.[73-74,79-80,85-87]

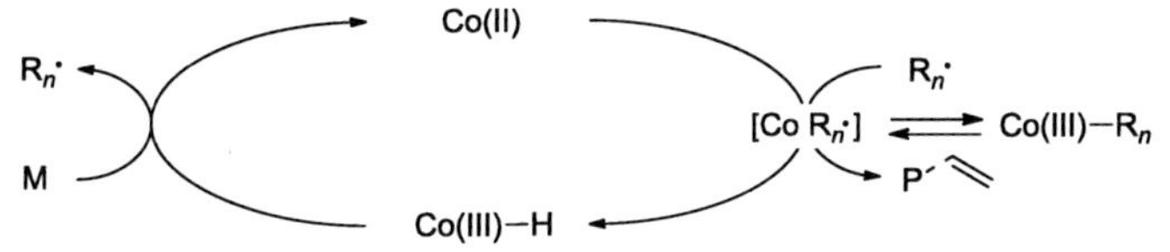

Scheme 3-7. Mechanism for the CCT polymerization mediated by a cobaloxime catalyst.[88]

The mechanism of CCT polymerization follows a two step radical process. In a first step the polymerization is started by a conventional initiation process. The obtained radical, represented by $R_n\cdot$, further reacts with the chain transfer agent. The cobaloxime oxidation state +II is increased to +III through the formation of a Co-H bond via hydrogen atom abstraction by the radical. For α-methyl monomers the ω-functionalized polymer is finally obtained. The Co-C bond formation – the side reaction of monomers without an α-methyl group – between catalyst and monomer is shown on the right hand side of Scheme 3-7.[89] The residual Co(III) complex is regenerated through the interaction with a monomer starting a radical polymer chain. A more detailed depiction of the hydrogen atom abstraction can be seen in Scheme 3-8.

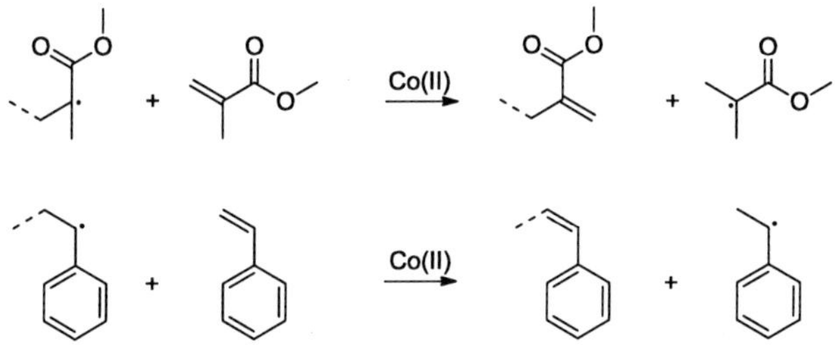

Scheme 3-8. Catalytic chain transfer of MMA and styrene. Mechanistic details for monomers with and without α-methyl group.

In the case of monomers bearing α-methyl groups, the abstraction of the hydrogen atom occurs from the α-methyl group to the propagating radical. As an example, the CCT polymerization of MMA is shown in the top reaction of Scheme 3-8. The terminus of the resulting macromonomer features a geminal double bond with a 1,1'-disubstitution.[77-79,90] In contrast to α-methyl containing monomers, the transfer of the hydrogen proton is achieved from the backbone of the propagating radical for vinylic monomers, i.e., styrene (as depicted in the bottom of Scheme 3-8), resulting in a vicinal endgroup with a 1,2-disubstituation.[79,81] The α-endgroup of the obtained macromonomers is represented by a hydrogen atom in both cases.

There are additional synthetic routes described in the literature to obtain macromonomers via CCT polymerization. Styrene based macromonomers have been synthesized via the copolymerization of styrene with a CCT derived dimeric precursor.[91] Additionally, Chiu *et al.* synthesized *n*-butyl acrylate macromonomers with either an α-methyl styrene (AMS) or benzyl methacrylate (BzMA) terminus via CCT polymerization,[92] evidenced by ESI-MS, ^{1}H- and ^{13}C-NMR spectroscopy characterization.

Further, Davis and colleagues synthesized ω-unsaturated polymers via a RAFT and CCT polymerization sequence.[93] Initially, poly(butyl methacrylate) and poly(methyl methacrylate) (pMMA) were synthesized via RAFT polymerization using dithiobenzoates as chain transfer agents. In a second step, the RAFT generated polymers were subjected to a CCT polymerization, employing the COBF catalyst complex in relatively high concentrations. The COBF concentration was consistent with dimer/trimer synthesis in a conventional CCT polymerization, however minimal polymerization occurred. Determination of the existing endgroup was carried out via ^{1}H-NMR which exhibited significant signals in the methacrylic region. Additionally, the elimination of the RAFT endgroup was analyzed via UV spectroscopy confirming a color loss of the polymer. Further, the purity of the polymer was determined using ESI-MS as characterization method. Copolymerization of the obtained macromonomers was then carried out with ethyl acrylate in a free radical polymerization to obtain graft polymers.

The CCT polymerization can readily be used for macromonomer synthesis, however, limited to monomers with an α-methyl group. Additionally, in the absence of the α-methyl group subsequent reaction steps need to be applied. However, one step synthetic approaches are preferable over multi-step sequences.

3.4.2 Thermolysis of RAFT Polymers

Macromonomers can also be synthesized via the thermolysis of RAFT polymers. RAFT generated polymers need to be heated to relatively elevated temperatures generating the desired unsaturated terminus. The Chugaev elimination chemistry at high temperature is well known for small molecules and was eventually adopted for polymers.[94-96]

In the literature several examples regarding the elimination of the RAFT endgroup exists.[4,97-102] Among them, Postma *et al.* reported in 2005 the thermolysis of polymers synthesized via RAFT polymerization,[103] whereby the thermal cleavage of the trithiocarbonate endgroup leads to an olefinic terminus. The type of olefin (substitution) is dependent on the nature of monomer used during the RAFT polymerization. The thermolysis of poly(styrene) affords a vicinal double bond with a 1,2-distubstitution, whereas acrylate based polymers, i.e., poly(*n*-butyl acrylate), result in geminal olefins with a 1,1'-disubstitution, as shown in Scheme 3-9. The thermolysis mechanism is similar to the proposed mechanism for degradation which is observed at temperatures above 300 °C.[98-100]

The cleavage of the RAFT endgroup requires elevated temperatures of 210 °C to 250 °C for poly(styrene)s and up to 370 °C for acrylate based polymers.

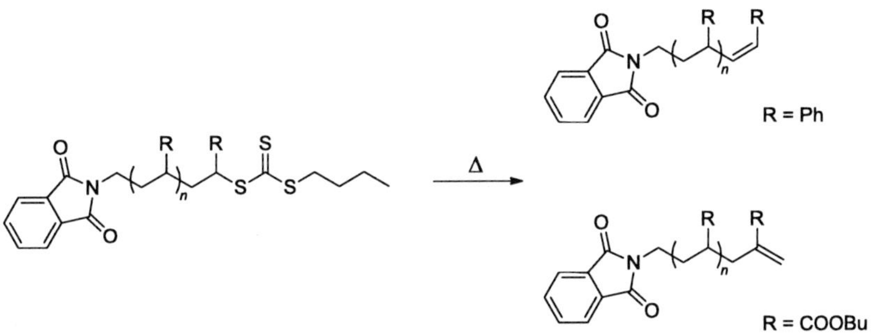

Scheme 3-9. Thermolysis of poly(styrene) and poly(*n*-butyl acrylate) resulting in macromonomers.[103]

Furthermore, Rizzardo and co-workers applied the thermolysis reaction to RAFT synthesized pMMA with dithiobenzoate and trithiocarbonate end-groups.[97] Elimination of the RAFT endgroup was observed at temperatures of 180 °C resulting in a terminal geminal olefin. The degradation of the RAFT endgroup was followed by thermogravimetric analysis, NMR and SEC. As already observed by Postma *et al.*, depending on the nature of the monomer, either vicinal or geminal double bonds are obtained from thermolysis.[103-104] This observation is explained through the mechanism proposed in Scheme 3-10.

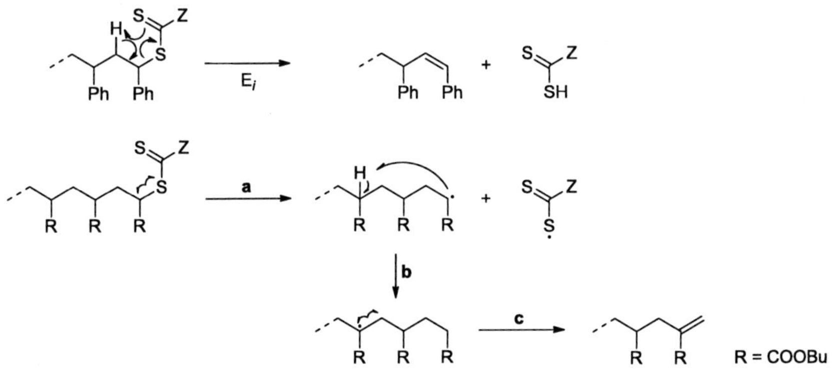

Scheme 3-10. Proposed mechanism for the elimination of the RAFT endgroup from poly(styrene) (top) and poly(*n*-butyl acrylate) (bottom). The reaction pathway includes concerted elimination (E_i) in the case of poly(styrene)s and homolysis of the C-S bond (a) followed by backbiting (b) and β-scission (c) in the case of poly(acrylate)s.

The mechanism in the upper half of Scheme 3-10 reflects the elimination via a concerted transition state which occurs for styrenic polymers.[4,103-104] In contrast to the Chugaev elimination of poly(styrene)s, the elimination of poly(acrylate)s follows an alternative reaction pathway. Firstly, homolysis of the C-S bond occurs followed by backbiting and β-scission (bottom Scheme 3-10). The unsaturated polymer obtained via RAFT thermolysis of styrenic polymers contains a vicinal double bond, however RAFT thermolysis of poly(acrylate)s results in a geminal terminus.

The cleavage of the RAFT endgroup by thermal degradation represents a facile route to obtain macromonomers. However, the thermolysis of RAFT polymers represents a two-step approach and the polymer and additional functionalities must be stable under these harsh conditions.

3.4.3 Vinyl-containing Initiators and Post Functionalization

The application of controlled polymerization protocols to macromonomer formation achieves defined polymer architectures with narrow polydispersities. Basically, two synthetic pathways are described in literature. The olefinic terminus – the characteristic feature of a macromonomer – can either be introduced by initiators containing a polymerizable group or by post functionalization of the obtained macromolecules.[22]

The former synthetic strategy was applied by Shen *et al.*[105-106] Vinyl containing initiators have been synthesized by ATRP initiation of different monomers. The obtained ATRP polymers featured a polymerizable endgroup. However, the polymerizable moiety of the initiator potentially acts as a monomer during the ATRP process resulting in branching. The authors showed that the polymerization of the vinyl terminus of the initiator does not occur below 80 % conversion. By terminating the ATRP at an early stage, macromonomers were obtained. Likewise, RAFT polymerization was initiated by a polymerizable RAFT agent. Öztürk *et al.* synthesized a RAFT agent with a polymerizable terminus at the R group.[107] The RAFT process, however, did not result in macromonomers but rather in highly branched and crosslinked polymer architectures.

The drawback of potential branching due to polymerization of the polymerizable terminus of the initiator requires additional synthetic strategies. Ferrari *et al.* circumvented the branching and applied ring-opening polymerization (ROP) to macromonomer formation.[108] The synthesis of the macromonomer is depicted in Scheme 3-11. The ROP process was initiated by hydroxyethyl methacrylate featuring a methacrylic functionality.

Scheme 3-11. Macromonomer formation through ROP initiators bearing a polymerizable group. Ring-opening polymerization initiated by hydroxyethyl methacrylate (HEMA).[108]

The latter synthetic strategy is based upon the post functionalization of macromolecules. The polymerizable group, which makes a macromonomer, is introduced by a further transformation. An abstract of several strategies is depicted in Scheme 3-12. Among others, Schön *et al.* synthesized ATRP polymers with a functional initiator (Scheme 3-12 (a)).[109] An excess of ligand in ATRP polymerization provided halogen free polymers with a terminal hydroxyl functionality via a hydrogen atom transfer. The subsequent esterification of the OH moiety with methacryloyl chloride yielded well-defined macromonomers.

Modular ligation via *click* chemistry was utilized by Vogt *et al.* (the concept of *click* chemistry will be discussed in Chapter 7).[110] The macromonomers are obtained through nucleophilic substitution of the bromine endgroup of the ATRP polymers with an azide by further *click* reaction with an alkyne (Scheme 3-12 (b)). Similar to this method, Topham *et al.* synthesized macromonomers through ATRP with an azide-containing initiator (Scheme 3-12 (c)) which makes a substitution step redundant.[111]

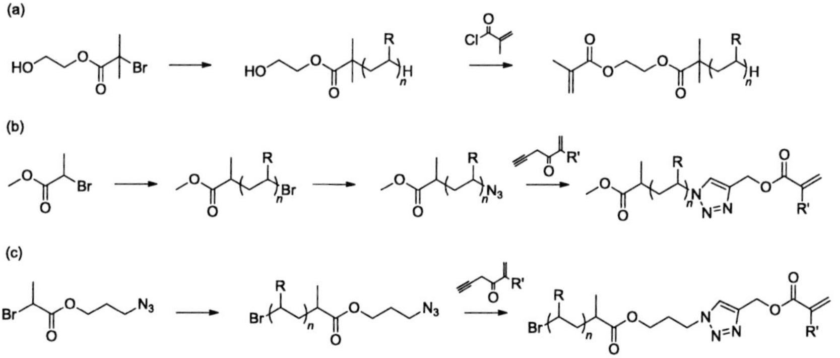

Scheme 3-12. Macromonomer formation through post-functionalization of polymers applying (a) esterification,[109] (b) nucleophilic substitution followed by *click* chemistry[110] and (c) *click* chemistry.[111]

4

Chromatography of Polymers

4.1 Separation Principles in Chromatography

The determination of the fundamental physical and chemical properties of macromolecules such as molecular weight, polydispersity and chemical heterogeneity is very important as it is these characteristics that are reflected in the macroscopic properties of the polymeric materials. In general, polymers are not monodisperse with a specific molecular weight but rather an assembly of different molecular weight macromolecules with a specific polydispersity. With regard to these characteristics, chromatography – as one classical characterization method – is an important source of analytical information. The separation of a polymer sample occurs on a column set which is the core component of any chromatography system. These columns are equipped with a porous substrate – the stationary phase – which, upon interaction with the analyte, affects separation. Each chromatography system runs on a specific solvent which solubilizes the polymer to be analyzed. This so-called mobile phase swells the stationary phase and serves as the vehicle by which the polymer analyte is eluted through the column. For sample analysis a specific quantity of polymeric material, generally 2 to 10 $mg \cdot mL^{-1}$, is injected into the columns and analyzed according to a specific separation mode. The retention time of an analyte, i.e., the elution volume, is influenced by the change in the Gibbs free energy ΔG. The separation of a polymer is affected by thermodynamic processes – enthalpic or entropic changes – which underpin the mode of separation.[112]

$$\frac{(V_e\text{-}V_m)}{V_s} = \exp\left(-\frac{\Delta G}{RT}\right) = \exp\left(-\frac{\Delta H}{RT} + \frac{\Delta S}{R}\right) \tag{4-1}$$

The change in either adsorption enthalpy ΔH or entropy ΔS of the polymer sample is influenced by the interaction of the solute with the mobile and stationary phase, which is associated with the elution volume V_e, the volume of the mobile phase V_m (dead volume of the column) and the volume of the stationary phase V_s.

If the separation of the sample is governed purely by entropic factors, ideal size exclusion chromatography (SEC) is observed. The injected analyte permeates the porous substrate of the stationary phase and this interaction between the analyte and the column material causes a change in entropy. The ideal SEC is depicted by the dotted straight line on the left hand side of Figure 4-1. The visualization of log(M) as a function of the elution volume V_e shows a linear decrease of molecular weight with increasing retention time, in other words large molecules elute prior to small molecules. Nevertheless, ideal SEC is never observed due to additional enthalpy changes through interaction of the analyte with the mobile phase. The 'real' SEC shows non-linearity with limitations in separation both at a lower molecular weight M_l ($V_e = V_i$) and at an upper molecular weight M_u ($V_e = V_m$) limit.

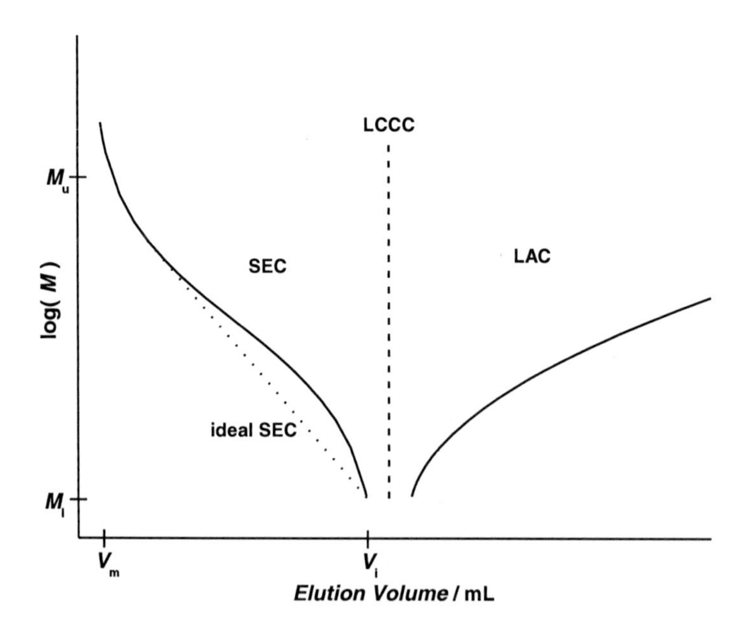

Figure 4-1. Seperation modes in liquid chromatography of polymers. The molecular weight as a function of the elution volume of the polymer sample. Depicted are (ideal) size exclusion chromatography (SEC), liquid chromatography at critical conditions (LCCC) and liquid adsorption chromatography (LAC).

On the contrary, separation via liquid adsorption chromatography (LAC) is driven by purely enthalpic interactions. The change in enthalpy is caused by interactions of the repeat units or functionality of the analyte with the stationary phase. During this process small molecules, i.e., minor quantities of repeat units and functionalities, elute faster than larger molecules which illustrates a reverse relation between log(M) and V_e (right hand side of Figure 4-1).

The special case in which the enthalpic and entropic effects are deliberately counterbalanced is referred to as liquid chromatography at critical conditions (LCCC). For a series of analytes of uniform chemical heterogeneity the LCCC separation is located between the SEC and LAC mode (dashed straight line Figure 4-1) and the log(M) equals a certain elution volume V_e independent of molecular mass. The critical conditions of chromatography need to be individually determined for a specific polymer type under special operating conditions including temperature, applied columns and the solvent mixture (solvent/non-solvent) of the mobile phase.

Hereinafter, both characterization methods – SEC and LCCC – are described in more detail as they constitute the fundamental analytical methods utilized in the present thesis.

4.2 Size Exclusion Chromatography (SEC)

In size exclusion chromatography (SEC) – also referred to as gel permeation chromatography (GPC) – the molecular weight and the polydispersity of a polymer can be determined. Most often, polymers are characterized through their number-average molecular weight M_n (4-2) or the weight-average molecular weight M_w (4-3), which are defined as follows:

$$M_n = \frac{\sum N_i \cdot M_i}{\sum N_i} \qquad (4\text{-}2)$$

$$M_w = \frac{\sum N_i \cdot M_i^2}{\sum N_i \cdot M_i} \qquad (4\text{-}3)$$

A polymer is properly characterized via SEC when at least two molecular weight averages, e.g., the above mentioned M_n and M_w, are determined. The polydispersity index (*PDI*) which describes the width of a distribution can then be calculated by equation (4-4).

$$PDI = \frac{M_w}{M_n} \qquad (4\text{-}4)$$

For a monodisperse polymer the ratio M_w/M_n equals unity and increases with increasing width of the molecular weight distribution. However, a monodisperse polymer, i.e., all polymer chains with the exact same length, is impossible to synthesize by chain growth processes. The formation of polymers with low *PDI* values is approachable via living/controlled radical polymerization techniques such as reversible addition fragmentation chain transfer (RAFT) polymerization,[3,113] atom transfer radical polymerization (ATRP)[2] and nitroxide-mediated polymerization (NMP).[1]

A chromatography system operating in the SEC mode separates the molecules according to their molecular weight, or more accurately by their hydrodynamic volume. The porous column material, e.g., microporous beads of crosslinked poly(styrene), is swollen in the solvent and the analyte is passed through the column with a constant solvent flow rate. During this process, the permeation according to the size of the macromolecules leads to a separation of the polymer sample. Small molecules are decelerated due to higher penetration of the pores and consequently result in a high retention time. Higher molecular weight fractions elute earlier from the columns than low molecular weight molecules. Conventional SEC systems are equipped with specific detectors for evaluation of the eluting fractions. The detection can either be achieved by the use of concentration sensitive detectors, such as refractive index (RI), ultraviolet-visible (UV-Vis) or infra-red (IR) detection or through the use of molecular weight sensitive detectors, such as multi angle light scattering (MALS) and low angle laser light scattering (LALLS).

SEC is referred to as a relative characterization method which requires calibration with narrow distributed polymer standards for a quantitative analysis. Most often, commercially available polymer standards such as poly(styrene) and poly(methyl methacrylate) are used. A range of polymer standards with specific molecular weights are measured on a SEC system to obtain a polynomial fitting curve which correlates the elution volume with the molecular weight. This correlation can then be used for a quantitative analysis of polymers similar to the polymer standards used for the calibration. This limitation to similar polymer types is circumvented if the empiric Mark-Houwink (Kuhn-Mark-Houwink-Sakurada) equation is employed (4-5).

$$[\eta] = K\,M^{\alpha} \qquad (4\text{-}5)$$

The molecular weight M is empirically related to the intrinsic viscosity $[\eta]$. The factors K and α are specific for the polymer type, the solvent and temperature are collated – at least for the majority of polymers – in literature.[114] In 1967 Benoit and co-workers observed that all polymer types fit to one specific calibration curve.[115] This universal calibration was obtained by plotting $\log([\eta] \times M)$ against the elution volume V_e and is only valid for a specific column set, solvent and temperature. The obtained calibration curve is of the same shape as the SEC curve in Figure 4-1. The change in only one parameter requires a new calibration with narrowly distributed polymer standards. If the Mark Houwink parameters K and α are known for the measured polymer sample the molecular weight can be calculated from the universal calibration curve based on the following equations.

$$[\eta]_1 M_1 = [\eta]_2 M_2 \tag{4-6}$$

$$\log M_2 = \log\frac{K_1}{K_2}\frac{1}{1+\alpha_1}+\frac{1+\alpha_1}{1+\alpha_2}\log M_1 \tag{4-7}$$

The molecular weight M_1 of the calibration standard with the Mark-Houwink parameters K_1 and α_1 are directly connected to the molecular weight M_2 of the measured polymer if the parameters K_2 and α_2 are known. The application of a universal calibration in SEC represents an attractive method for molecular weight analysis if the Mark-Houwink parameters are known; however universal calibration does not offer the same accuracy as is available by absolute molecular weight determination with light scattering detectors.

4.3 Liquid Chromatography at Critical Conditions (LCCC)

The use of liquid chromatography at critical conditions (LCCC), as an additional characterization method, yields more detailed information regarding molecular weight and chemical heterogeneity.[116] In LCCC the molecules are separated by their chemical heterogeneity in the absence of size exclusion effects. At critical conditions, which are located between the SEC mode and LAC mode (refer to Figure 4-1), polymers with identical repeating units, functionalities and topology elute at the same retention time, independent of their hydrodynamic volume. The analysis of a block copolymer A-*block*-B at critical conditions of segment A leads to elution volumes related to the segment B, whereas the segment A is chromatographically invisible. This situation is inverted when the critical conditions of segment B are utilized. LCCC characterization thus leads to a more detailed visualization of the block building efficiency since the block segment of a specific block in a copolymer elutes at a different retention time to the other block. A LCCC dimension hyphenated with conventional SEC, i.e., two-dimensional or 2D chromatography, provides information on the chemical heterogeneity and size of the synthesized copolymers, respectively. In the literature LCCC as a straight forward characterization method, especially hyphenated with SEC, is employed by several research groups for a detailed analysis method.[1,112,117-128]

4.4 Determination of Critical Conditions

The LCCC measurements require the knowledge of the critical conditions which are related to a specific polymer type. A variety of combinations of column sets, either normal or reversed phase, temperature and solvent mixtures are conceivable for critical conditions. In literature a small selection of critical conditions is available.[112,119,127,129-132] However, more often critical conditions need to be determined for individual purposes.
For the LCCC characterization in this thesis, the critical conditions of poly(*n*-butyl acrylate) (pBA) and poly(ε-caprolactone) (pCL) have been determined. The critical conditions of pBA and pCL applied for copolymer analysis are summarized in Table 4-1.

Table 4-1. Summary of the critical conditions applied within this thesis. The critical conditions were found at 35 °C at a flow rate of 0.5 mL·min^{-1}. For detailed specifics refer to Chapter 10.

polymer type	column phase	solvent A	solvent B	A / B (v/v)
pBA	normal	THF	*n*-hexane	28.7 / 71.3
pBA	reversed	THF	MeOH	40 / 60
pCL	reversed (2×)	THF	MeOH	20 / 80

In the following, the general approach for the determination of critical conditions will be illustrated for pBA which were applied in Chapter 8 for copolymer characterization.

For evaluation of the critical conditions, polymer standards with a narrow polydispersity and high endgroup fidelity need to be synthesized. In an ideal case, the endgroup of the polymer standards are identical with the endgroups of the applied starting materials in block copolymer formation. In this work, the pBA standards have been synthesized via the RAFT technique and the resulting SEC elugrams are depicted in Figure 4-2.

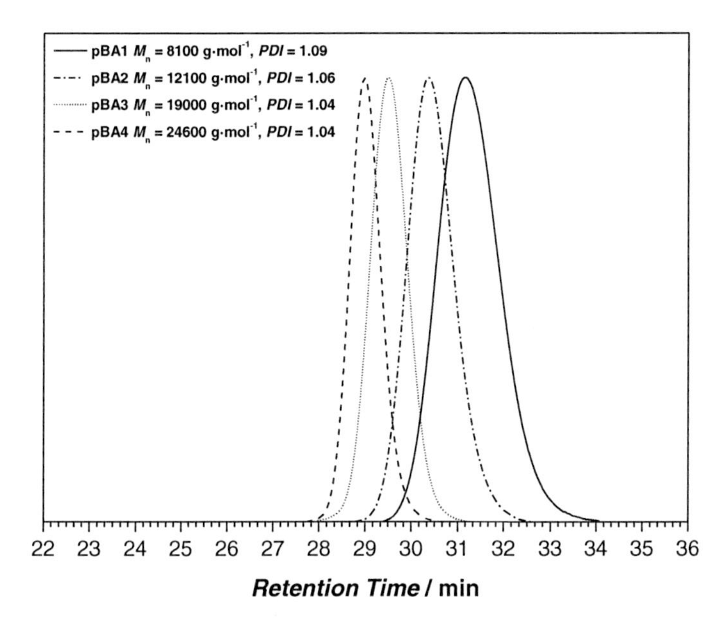

Figure 4-2. SEC elugrams of the poly(*n*-butyl acrylate) standards synthesized via the RAFT technique. The standards were applied to measure critical conditions for the LCCC measurements.

The molecular weights of the pBA standards ranged from M_n(pBA1) = 8100 g·mol^{-1} to M_n(pBA4) = 24600 g·mol^{-1} with a polydispersity of up to 1.09. For critical condition determination, these pBA standards were injected into a specific column set (in the present case a normal phase column) which can feature different modifications and pore sizes of the stationary phase. Initially, the pBA samples were measured with 100 % of a good solvent (here THF). The flow rate was set to 0.5 mL·min^{-1} for accelerated measurements, whereas slower flow rates could also be applied (down to 0.01 mL·min^{-1} for 2D SEC). At 100 % THF, the pBA standards eluted in SEC mode where the elution volume increases with the decrease in molecular weight, i.e., pBA4 elutes before pBA3, pBA2 and pBA1. Gradually a poor solvent (here: *n*-hexane) was added and the standards were remeasured. The samples were always prepared and injected with the solvent mixture currently run on the system.

The general progress towards the critical conditions can be visualized by plotting the number-average molecular weight M_n against the elution volume V_e (Figure 4-3). For each polymer at a specific solvent mixture the elution volume (taken from the peak maximum) is plotted against its molecular weight resulting in four (depending on the amount of standards) data points. The measurements at 100 % THF are represented by the squares on the far left hand side of Figure 4-3 and are connected with a straight line to guide the reader's eye. At a solvent mixture 40/60 (v/v), represented by the triangles, the peak maxima are slightly shifted to higher elution volumes. With the approximation to the critical conditions the peak maxima of the pBA standards are shifted to higher elution volume which can be clearly seen in Figure 4-3. At a solvent mixture composition in which the separation occurs in LAC mode the polymer standards are adsorbed onto the column and the elution volume increases with increasing molecular weight. The experimental data shown in Figure 4-3 clearly evidence the theoretical reflections visualized in Figure 4-1.

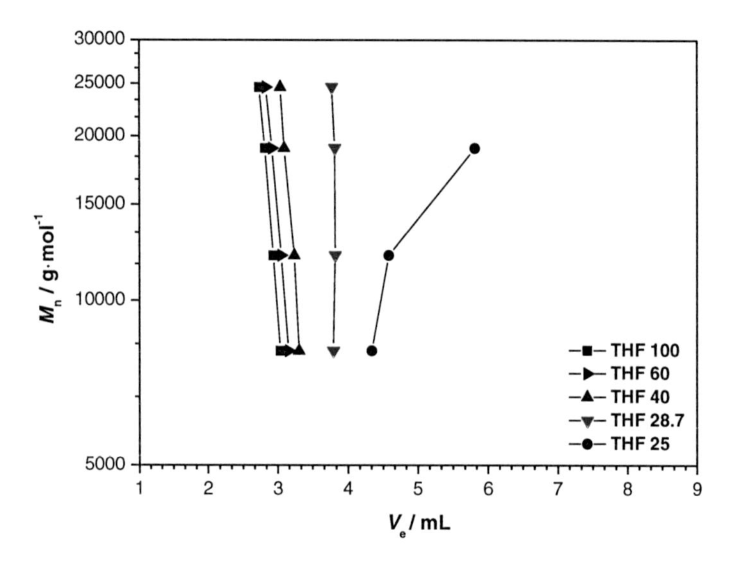

Figure 4-3. Elution volume of poly(*n*-butyl acrylate) standards versus molecular weight at different solvent mixtures of the mobile phase on a normal phase column.

Another manner of representation, contrary to the graphical view in Figure 4-3, involves the relation of the retention factor k ($k = (V_e - V_m)/V_m$) to the volume percentage of THF which is shown in Figure 4-4. For each solvent mixture, four k values are obtained from the peak maxima of the polymer standards. The volume of the mobile phase V_m is obtained from the measurement of a high molecular weight polymer, e.g., $M_n = 10^6$ g·mol^{-1} which does not penetrate the porous column material and elutes unimpeded. The retention factor k, i.e., the elution volume V_e, of the pBA standards increases with further addition of *n*-hexane (from the right to the left hand side of Figure 4-4). In addition to the increasing elution volume, the distance of the peak maxima of pBA1 to pBA4 is reduced as clearly evidenced by the values moving together (e.g., compare the measurements at 100 % and 40 % THF).

At one characteristic point – the critical conditions – all standards elute at the same elution volume independent of their size (Figure 4-5).

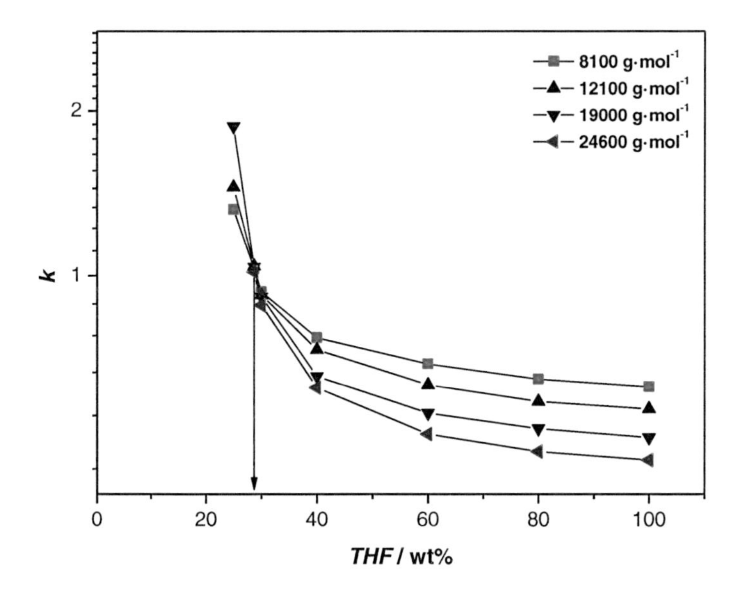

Figure 4-4. Determination of the critical conditions of poly(*n*-butyl acrylate) on a normal phase column. Critical conditions are found to be at a solvent mixture of THF/*n*-hexane 28.7/71.3 (illustrated by the arrow).

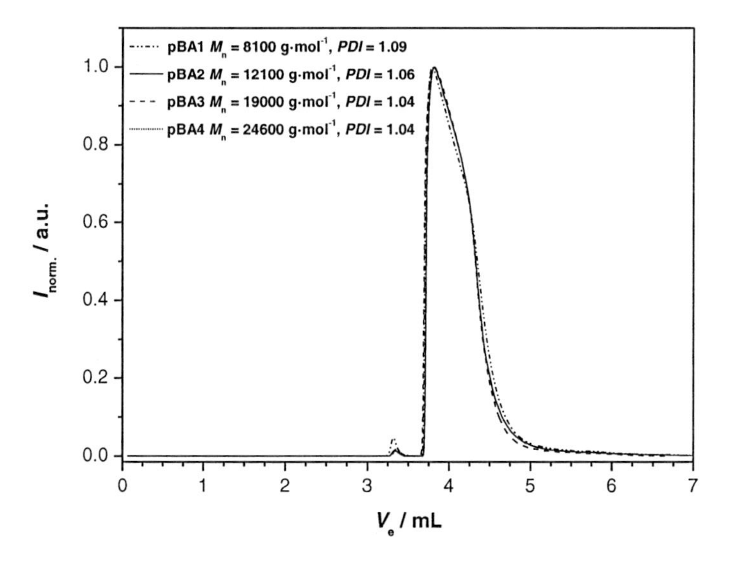

Figure 4-5. poly(*n*-butyl acrylate) samples with different molecular weight at critical conditions on a normal phase column with a solvent mixture of THF/*n*-hexane 28.7/71.3.

The critical conditions of pBA are represented by the red triangles at 28.7 % THF in Figure 4-3 which shows a straight line at a defined elution volume. In Figure 4-4 the defined elution volume at the critical conditions is reflected by the point at which the different *k* values superimpose each other.

The critical conditions of pCL have been determined in an identical fashion with pCL polymer standards synthesized via organo-catalyzed ring-opening polymerization (ROP).

4.5 LCCC Application in Copolymer Characterization

In the literature LCCC as a powerful characterization method, especially hyphenated with SEC, has been employed by several research groups for a detailed analysis method of (block) copolymers. [1,112,117-128] It allows for an evaluation of block copolymer formation efficiency and additionally can clarify the molecular weight and chemical composition.
For example, Matyjaszewski and co-workers subjected poly(methyl acrylate)-*b*-poly(styrene) linear and star shaped block copolymers synthesized via (activator generated by electron transfer) (AGET) atom transfer radical polymerization (ATRP) to LCCC.[118,120,133] The obtained elugrams from the 2D analysis showed that pure block copolymers were obtained without any residual macroinitiator or homopolymers which could not be accurately determined by conventional SEC alone. The quantitative reaction of the macroinitiators in ATRP was clearly visualized by the use of LCCC.

Furthermore, living anionic polymerization has been utilized by Falkenhagen, Müller and colleagues for poly(methyl methacrylate)-*b*-poly(*tert*-butyl methacrylate) block copolymer synthesis and subjected to LCCC analysis on both normal and reversed phase columns.[112] The copolymers have been completely characterized on a normal phase system for poly(*tert*-butyl methacrylate) and a reversed phase system for poly(methyl methacrylate). The analysis technique – investigating both polymer types of the block – offers simple and fast access to identical results which would be obtained from 2D chromatography.
The application of LCCC, or rather 2D chromatography, was applied to determine the efficiency of rapid modular conjugation by Inglis *et al.*[119] The chromatographic analysis showed quantitative conjugation efficiency comparable to theoretical reflections. Likewise, Wong *et al.*[134] evidenced the formation of star block copolymers via 2D chromatography. Recently, Schmid *et al.* used this chromatography technique for the characterization of poly(styrene)-*b*-poly(ε-caprolactone) block copolymers synthesized via combining RAFT polymerization with ring-opening polymerization (ROP).[126-127] The benefit of the 2D measurement was clearly illustrated by a reference measurement. A sample with non-quantitative block copolymer formation was compared to efficient block copolymer synthesis (Figure 4-6).

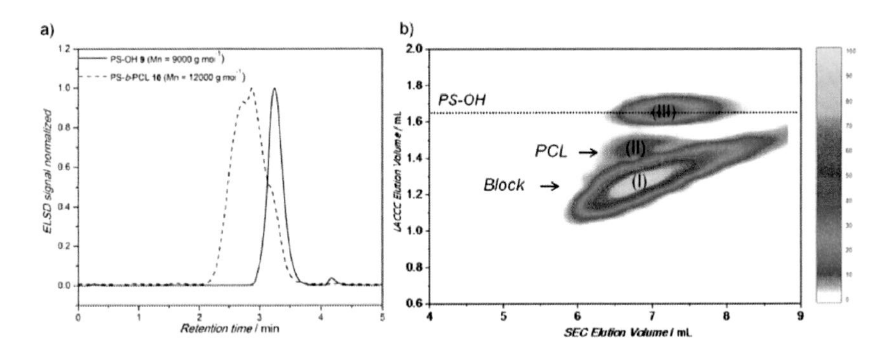

Figure 4-6. LCCC elugram (a) and 2D contour plot (b) of poly(styrene)-*b*-poly(ε-caprolactone) copolymer at critical conditions of poly(styrene) with residual homopolymer (OH terminated poly(styrene)). Adapted with permission from Schmid C.; Falkenhagen J.; Barner-Kowollik C., Journal of Polymer Science Part A: Polymer Chemistry.[126] Copyright (2011) John Wiley & Sons.

In contrast to the efficient block copolymer formation, which showed detached elugrams with minor amounts of residual macroinitiator, the reference measurement displayed a multimodal distribution with three contributions implying the presence of residual macroinitiator, block copolymer and water initiated pCL homopolymers. The authors clearly pointed out the strength of LCCC as a useful methodology in block copolymer characterization.

The characterization of block copolymers and the associated construction efficiency is of significant importance in the present thesis. For this reason, the synthesized block copolymers prepared either via copolymerization techniques (Chapter 8) or from macroinitiators (Chapter 9) have been subjected to LCCC.

5

Mass Spectrometry of Polymers

5.1 Introduction

Mass spectrometry (MS) is widely used for elemental composition analysis and allows for the elucidation of the molecular structure of the analyte.[135-136] This characterization technique provides the mass-to-charge ratio (*m/z*) of gaseous ions of the analyte and is applicable to a variety of components such as organic compounds, biopolymers and synthetic polymers.[137-138] Mass spectrometry is well suited for the characterization of polymers due to its high sensitivity, which allows the analysis of low quantities of material; high selectivity (low quantities of specific species can be detected) and rapid measurements.[139] Figure 5-1 outlines the general components of a mass spectrometer which consists of a sample inlet for analyte injection, an ion source where the ionization process occurs, separation of the analyte via *m/z* in the mass analyzer and the final detection. The computer-generated mass spectrum is then available for detailed interpretation. Parts of the mass spectrometer operate under high vacuum conditions to avoid collision of gaseous sample ions with background gases.

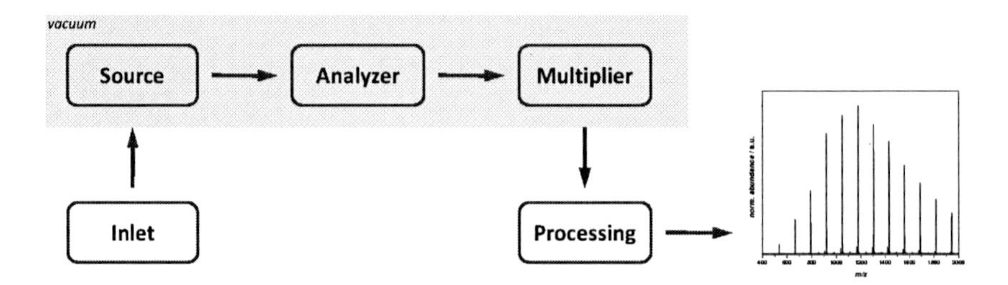

Figure 5-1. Basic instrumental setup of a mass spectrometer including ion source, mass analyzer, ion detection and data processing.

At first, the analyte is vaporized and ionized allowing the analysis and detection of the generated gaseous ions according to their *m/z* ratios. These include either molecular ions M^+ or quasimolecular ions, e.g., $[M+Na]^+$, depending on the applied ionization method. The basic ionization methods include 'hard' ionization, yielding fragmentation of the polymer and 'soft' ionization, with minor or no fragmentation observed. In the case that fragmentation occurs, the fragments provide information about endgroups and the general sequence of the monomer units of the polymer. However, soft ionization represents the method of choice for polymer characterization for primary structure determination. Fragmentation could subsequently be induced via collision-induced dissociation (CID)[140] or multi-dimensional mass spectrometry experiments (MS^n) if needed. Typical ionization methods are fast atom bombardment (FAB),[141-143] electron (EI), chemical (CI) and field (FI) ionization for 'hard' ionization and soft ionization techniques such as, electrospray ionization (ESI),[144] matrix-assisted laser desorption (MALDI),[145] atmospheric pressure chemical ionization (APCI)[146-148] and field desorption (FD) for 'soft' ionization. The mass analyzer then separates the analyte in the gas phase according to the *m/z* ratios of the various constituents.

In the following, the utilized components of the mass spectrometric setup used in the present thesis will be discussed in detail.

5.2 Ionization Methods

The application of mass spectrometry in polymer chemistry predominantly benefited from the progress in the development of ionization methods.[149] Polymers are most often high molecular weight materials and hard to vaporize. The focus on soft ionization methods, which prevent fragmentation during the ionization process, greatly simplified the interpretation of the obtained mass spectra. Commonly, electrospray ionization (ESI) as well as matrix-assisted laser desorption (MALDI) are used for mass spectrometry on polymers.[150-152]

5.2.1 Electrospray Ionization (ESI)

The electrospray ionization (ESI) technique first engaged attention in the late 1960s by the pioneering work of Dole *et al.*[153-154] Dole and his co-workers showed that the electrospray technique produces macroions in the gaseous state by simple electrospraying of a dilute polymer solution. A high electrical field is applied for generating a continuous charged spray. However, multiply charged ions did not allow for the accurate determination of molecular weight. Further development in the field of ESI was achieved by Fenn and co-workers in the mid 1980s.[155-156] They evidenced the ionization of protein molecules up to 4×10^4 $g \cdot mol^{-1}$ without detectable fragmentation of the generated macroions.[157-158] The tremendous enhancement of the ionization process towards high molecular weight (bio)polymers by Fenn was finally rewarded with the Nobel Prize for Chemistry in 2002.[159]

During the ESI process, the ions required for m/z detection are generated by the application of a strong electric field.[160-161] A basic illustration of the ESI mechanism is depicted in Figure 5-2.

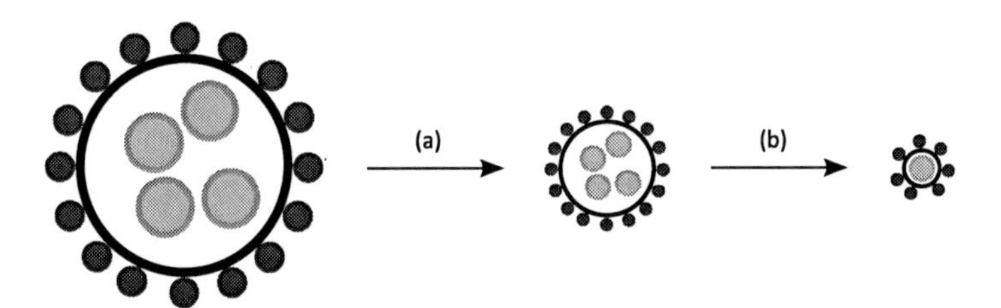

Figure 5-2. Basic electrospray ionization mechanism occurring in an electric field (from spray capillary to skimmer). The evaporation of the solvent (a) is followed by a series of Coulomb explosions (b) up to isolated gaseous ions. The analyte is represented in light gray, the positive charges in dark gray.

The nebulization of the analyte solution occurs at ambient temperature and atmospheric pressure through the application of high voltage towards the ESI needle. Due to charge repulsion on the surface of the capillary a so-called Taylor cone is formed. A continuous injection of the analyte solution through a capillary produces electrically charged droplets in the spray chamber. The charge of the ions depends on the polarity of the electric field, a positive potential generates positively charged ions or quasimolecular ions and vice versa for a negative potential. Often, salts are added to support the ionization process such as Na^+, which induces multiply charged (n) quasimolecular ions $[M+nNa]^{n+}$. Solvent evaporation further occurs via a heated 'transfer' capillary or contralateral flow of a nitrogen 'support' gas. However, the desolvatisation is also promoted by a coaxial flow of a nebulizing gas (most commonly nitrogen). The generated quasimolecular ions are ejected into the vacuum via electrostatic repulsion and further analyzed with respect to their *m/zs*.

Up to now, the mechanistic details of the ionization process in the ESI technique from charged droplets to gas phase ions are not yet clarified. [162-165] The suggested theory by Dole is referred to as a charged-residual mechanism (CRM),[13] whereby the electric charge of the droplets increases with increasing desolvatisation.[166] Once the Rayleigh limit is reached, where the electrostatic repulsion of the charges is greater than the surface tension of the droplets, the droplets explode in a so-called Coulomb explosion forming smaller droplets.

The process of successive Coulomb explosions is repeated until a single molecular ion is generated. Conversely, according to the ion-emission mechanism (IEM) by Iribarne and Thomas, the analyte ions' are generated from the charged droplets, which are also generated from continuous Coulomb explosions, via field-assisted desorption or ion evaporation.[167-168] The gaseous analyte ions are directly emitted from the droplets due to high electrical fields on the droplet surface. A combination of both CRM and IEM for the ionization process in ESI was suggested by Hogan *et al.*[169-170]

The ESI process generates ions of non-volatile polymers without any fragmentation. The formed ions are subsequently transferred to the mass analyzer for *m/z* analysis.

5.2.2 Matrix-assisted Laser Desorption Ionization (MALDI)

Contrary to the ESI technique, matrix-assisted laser desorption ionization (MALDI) represents ionization from the solid state. The utilization of MALDI as an ionization method goes back to the work of Karas and Hillenkamp in the late 1980s.[171-172] MALDI as a tool for polymer ionization received great interest since Karas illustrated non-fragmented ionization of proteins in a high molecular weight range (up to 10^5 $g{\cdot}mol^{-1}$).[173] The basic characteristic of MALDI is soft desorption.[174] For polymer characterization via MALDI, the analyte is dissolved and embedded into an organic matrix compound which incorporates the molecules. Typical matrices for MALDI mass spectrometry such as 2,5-dihydroxybenzoic acid (DHB), α-cyano-4-hydroxycinnamic acid (α-CHCA) and 2-(4-hydroxyphenylazo)-benzoic acid (HABA) are depicted in Figure 5-3.

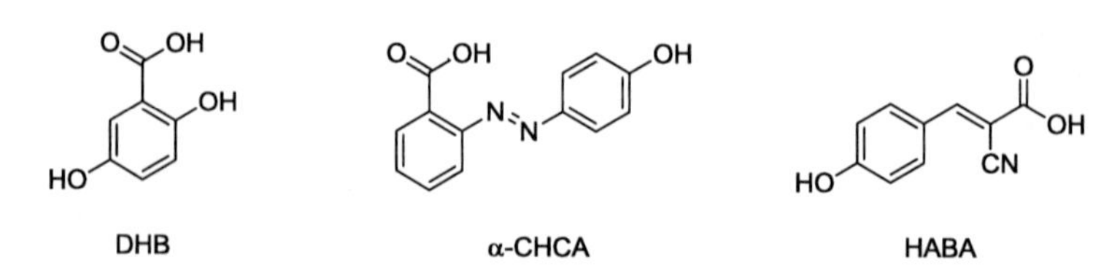

Figure 5-3. Chemical structures of matrices used in MALDI mass spectrometry.

The sample preparation often also includes the addition of an ionization agent, i.e., a metal ion salt, to encourage the ionization process. The sample solution is then applied to a target plate and the residual solvent is evaporated leaving a crystalline sample to be exposed to the MALDI process. The ionization process is then initiated by irradiation with laser light. The accessory matrix absorbs at the wave length of the emitted laser light and thus converts the incoming energy into heat. Thermal desorption of the analyte followed by desolvation and H^+/X^+ transfer to the analyte (X represents the ionization agent) generates the required gaseous ions for *m*/*z* analysis and detection.[171-172,175-177] Even to date, the detailed mechanism of the MALDI process is barley understood.[136] Similary to ESI, the transfer to the gas phase occurs without fragmentation. Multiply charged ions are observed, however to a minor extent compared to ESI.

5.3 Mass Analyzer

The mass analyzer – as a part of the mass spectrometer – sorts the generated gaseous ions according to their *m*/*z* values in either space or time.[178] The design of the mass analyzer should be carefully chosen to comply with the analytical problem, i.e., covered *m*/*z* range, resolution power, mass accuracy of the measurements, sensitivity and the opportunity to perform tandem measurements (MS/MS). Varying demands can be fulfilled with mass analyzers, such as sector (magnetic and/or electric) mass analyzers, quadrupole mass filters, quadrupole ion traps, time-of-flight (ToF) mass analyzers, orbitrap and Fourier Transform Ion Cyclotron Resonance (FTICR) mass analyzers. The *m*/*z* range of a ToF mass analyzer (up to 10^6) is found to be the upper limit.[150] The accuracy of the mass analyzer represents a crucial factor of accurate mass assignment and structural determination. The best performance in terms of accuracy can be achieved with a FTICR or orbitrap mass analyzer (< 5 ppm). Upon high mass accuracy the assignment of a specific signal, i.e., *m*/*z* value, is reduced to few imaginable structures and thus features a higher probability of accuracy.

The appearance of a specific peak is dictated by the resolution of the applied mass analyzer. A peak appearing at a specific *m*/*z* value can either be unresolved or feature an isotopic pattern. As an example, the carbon as a main element of a polymer backbone contains the stable isotope ^{13}C (1.1 %) in addition to ^{12}C (98.9 %). Consequently, for every peak associated with a carbon product appearing at some value of *m*/*z*, an additional smaller peak is observed at *m*/*z* +1. Several atoms such as bromine or silver feature a unique isotopic pattern which gives detailed information about the chemical structures. Likewise, with regard to the accuracy, the FTICR mass analyzer represents the instrument of choice for a high resolution power. The use of a FTICR, however, is limited due to the high costs associated with its purchase.

In the following, the quadrupole ion trap and ToF mass analyzer will be discussed in detail as the mass analyzers applied for this work.

5.3.1 Quadrupole Ion Trap

A quadrupole mass filter consists of four parallel circular hyperbolic rods with alternating polarities through the application of appropriate voltages of the same magnitude. The voltage features direct current (DC) as well as a radio frequency (rf) component. Gaseous ions are accelerated into the center of the four rod arrangement and subjected to the applied DC and rf voltages defining the fate of the ions.[179] At a constant DC/rf ratio the voltage is changed to obtain stable or unstable trajectories for ions of a specific *m/z* value. The separation subsequently occurs by holding the different masses on stable trajectories.

A quadrupole ion trap can be considered to be a three dimensional quadrupole mass analyzer with a ring electrode and two end caps.[180] A typical quadrupole mass filter is illustrated in Figure 5-4.

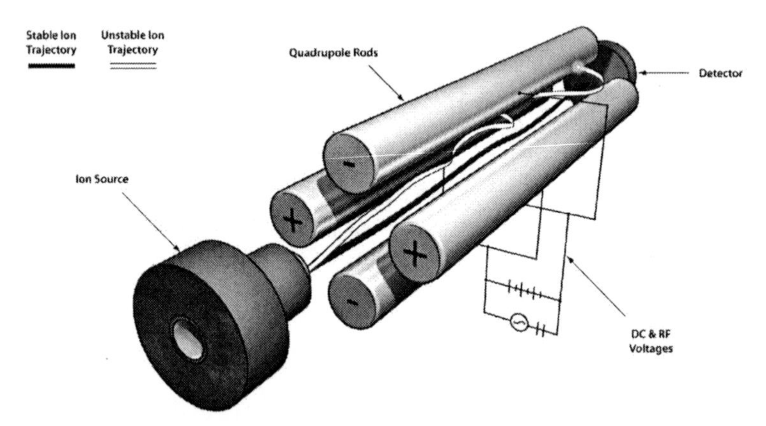

Figure 5-4. An illustration of the basic components of a quadrupole mass filter system. Adapted with permission from Hart-Smith, G.; Blanksby, S. J., Mass Analysis in *Mass Spectrometry in Polymer* Chemistry.[139] Copyright (2011) John Wiley & Sons.

Low amplitude rf is applied to the ring electrode whereas the end caps of the quadrupole ion trap are grounded. The gaseous ions are injected into the trap from the ESI device where the ions experience repulsive forces. To prevent collision of the gaseous ions of an analyzed sample which would lead to a premature departure the trap, the gas atmosphere inside the quadrupole is filled with Helium gas. The Helium gas absorbs the collision energy and keeps the gaseous ions inside the trap. The rf potential is then scanned to eject the ions from the trap towards the detector with increasing *m/zs*.[181-182]

The quadrupole ion trap is a convenient mass analyzer to accomplish multi-dimensional mass spectrometry. For a MS/MS experiment the ions of a specific *m/z* value are trapped inside the quadrupole by applying a DC voltage to the ring electrode and subsequently increasing rf voltage (first dimension). For the actual MS/MS experiment an alternating voltage is applied to the end caps which induces fragmentation (i.e., formation of additional ions). The obtained fragments give an indication about the endgroups or the general structure of the polymer backbone. The mass analysis of the ions occurs via scanning the rf voltage as described above.

5.3.2 Time-of-Flight (ToF)

In a time-of-flight (ToF) mass analyzer the gaseous ions are separated in time by their associated *m/z*s according to their velocity.[183-184] For that to happen, the gaseous ions are accelerated within a fixed potential and analyzed by their flight time through a defined distance *d* until they hit the detector.[185] Each ion experiences the same kinetic energy at the start of the acceleration. The flight time *t* is provided by the velocity *v* of the ions and thus connected to the mass *m* through the relations $v = d/t$ and $E_{kin} = 0.5\, mv^2$; i.e., lower mass ions arrive earlier at the detector than higher mass ions. The *m/z* value can thus be calculated from equation (5-1) for a given distance *d* and flight time *t* with the elemental charge *e* and the applied potential *V*.

$$\frac{m}{z} = \frac{2eVt^2}{d^2} \tag{5-1}$$

In reality, however, ions of the same mass can feature different final velocities and thus different kinetic energies caused by acceleration from different locations within the instrument. Therefore, a linear ToF instrumentation features a relatively low resolution. An increase in resolution can be achieved via delayed extraction[186] or a reflectron.[187-188] The use of delayed extraction provides a time lag between the ionization process and the acceleration of the gaseous ions which results in their simultaneous approach to the detector. The latter improvement towards ToF analysis benefits from the retardation and reflection of ions. Ions of the same *m/z* values, yet different velocity drift into the reflectron with a specific electric field according to their kinetic energy. Faster ions drift deeper into the reflectron than slower ions and thus simultaneously leave the reflectron. The use of a ToF mass analyzer in combination with the MALDI ionization technique is a promising tool for polymer characterization up to 10^6 g·mol^{-1}.[189-190]

5.4 Detectors

The detector represents the final unit (before data processing occurs on a computer which represents the actual final component) utilized in a mass spectrometer. The obtained *m/z* values are converted to a measurable electrical signal. Usually, an electron or photon multiplier is used.[191] The mass spectrometer used in this work features an electron multiplier for ion detection and will be described in the following.

5.4.1 Electron Multiplier

Two forms of electron multipliers are employed for detection in mass spectrometry: discrete dynode electron multiplier and continuous (channel) dynode electron multiplier.[192-193] The performance of the mass spectrometer is substantially influenced by the efficiency of the ion detection and thus the efficiency of the electron multiplier. The functional principle of the electron multiplier is based on the so-called secondary electron emission.[194-195] When the incident single electron or ion reaches the surface of the electron multiplier, it causes the emission of secondary electrons from the semi-conducting surface material. The basic geometry of electron multipliers is depicted in Figure 5-2.

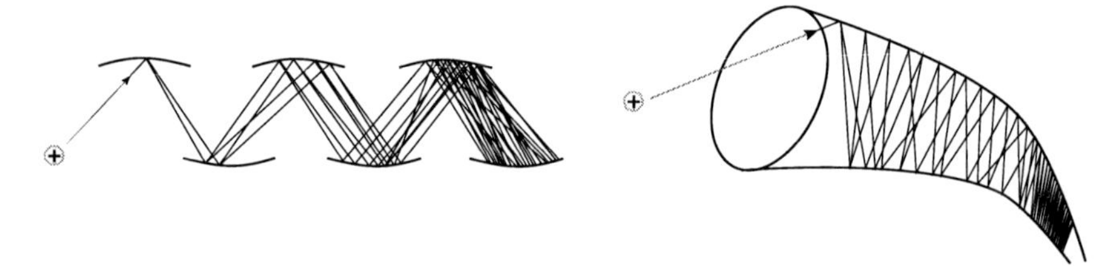

Figure 5-5. Basic mechanics of a discrete dynode electron multiplier (left) and a continous dynode electron multiplier (channeltron) (right).

The application of a high electrical potential accelerates the emitted secondary electron again to the surface and causes the emission of more secondary electrons. Through the repetition of this cascade collision process one incident electron undergoes amplification. The design of a discrete dynode electron multiplier is attributed to sequential multiplying spatially separated electrodes. Contrary to this, a continuous dynode structure – which represents the most widely used design – features funnel geometry and is commonly named 'channeltron'. Multiple continuous dynode electron multipliers arranged in parallel in a two dimensional geometry are referred to as microchannel plate or array detectors.[196] The repetition of the secondary electron emission typically occurs 10 to 20 times until the amplified electrons hit the anode, whereupon conventional amplifiers are installed in series. The obtained electrical signal provides the actual m/z values of the detected ions.

5.5 ESI-MS hyphenated with Size Exclusion Chromatography

The mass range for analysis of polymeric samples is defined by the mass analyzer applied in a mass spectrometer.[197] A direct infusion measurement where the ionization occurs directly after the sample injection is limited to $m/z = 2000$ for the linear quadrupole ion trap employed herein (in some cases up to $m/z = 4000$). The formation of multiply charged gaseous quasimolecular ions during the ESI process enables the analysis of higher molecular weight polymers outside of the m/z range due to decreasing m/z values for increasing attached ions.

However, mass spectrometry of polymers with higher molecular weights, i.e., above 10^5 g·mol^{-1}, is feasible by pre-separation of the injected polymer sample via size exclusion chromatography (SEC). The general instrumental setup of a mass spectrometer coupled to a SEC (SEC/ESI-MS) is illustrated in Figure 5-6.

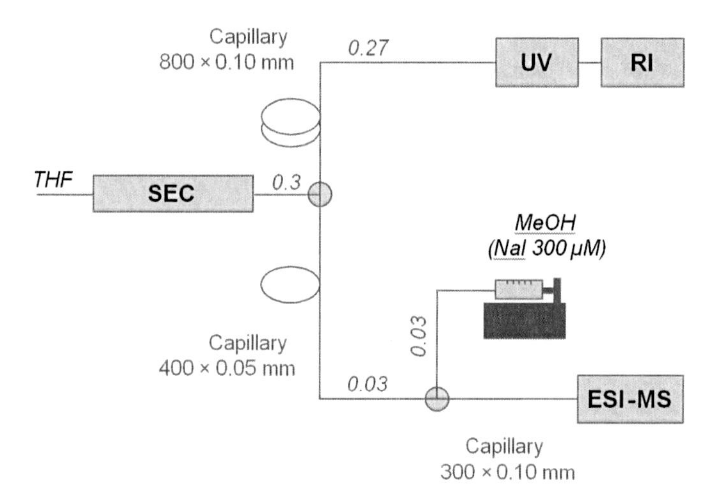

Figure 5-6. Schematic SEC/ESI-MS setup employed in this work. The numbers indicate flow rates in mL·min^{-1}. Adapted with permission from Gruendling, T; Guilhaus, M.; Barner-Kowollik, C. Analytical Chemistry.[197] Copyright (2008) American Chemical Society.

For SEC/ESI-MS analysis, the analyte is dissolved in an appropriate solvent, e.g., THF, and injected into a chromatography system. The separation then occurs in SEC mode, i.e., high molecular weight species elute prior to low molecular weight species. After the pre-separation of the analyte, the effluent is split into parallel flow pathways. Seven-eighths of the elution volume is directed to the pathway (top, Figure 5-6) giving rise to molecular weight determination via concentration sensitive RI and UV-detectors as in a conventional SEC instrumentation setup.[197] On the other hand, one-eighth of the elution volume is transferred to the mass spectrometer (bottom, Figure 5-6). The continuous addition of a sodium iodide solution in methanol prior to the injection to the ESI source improves the ionization efficiency and yields multiply charged species. Mass spectra are continuously recorded for each fraction eluting from the SEC columns.

5.6 Interpretation of Mass Spectra

Mass spectra of polymers differ from mass spectra obtained for organic substrates or biomolecules with a defined mass. For polymers with a specific polydispersity various chain lengths are present in the sample to be analyzed. Each chain length features a defined *m/z* value and is visible in the mass spectrum. A typical mass spectrum of *n*-butyl acrylate macromonomer is depicted in Figure 5-7.

It is clearly seen that the signals assigned to one species are repeated at a constant interval according to the *m/z* value of the repeat unit, the monomer. In the example above, the main distribution with the most abundant signals can be assigned to the main species $\mathbf{MM}^{\mathrm{H}}$ (macromonomer with a geminal double bond) as discussed earlier (Chapter 3). The ionization is supported by the addition of a sodium salt which results in a quasimolecular ion $[M+Na]^{+}$ and the attachment of sodium to the polymer species needs to be accounted for molecular mass determination. Uncharged species are not visible in mass spectrometry and neither be detected nor analyzed. However, additional to the species $\mathbf{MM}^{\mathrm{H}}$ found in the mass spectra, several other species can be found associated with one specific repeat unit.

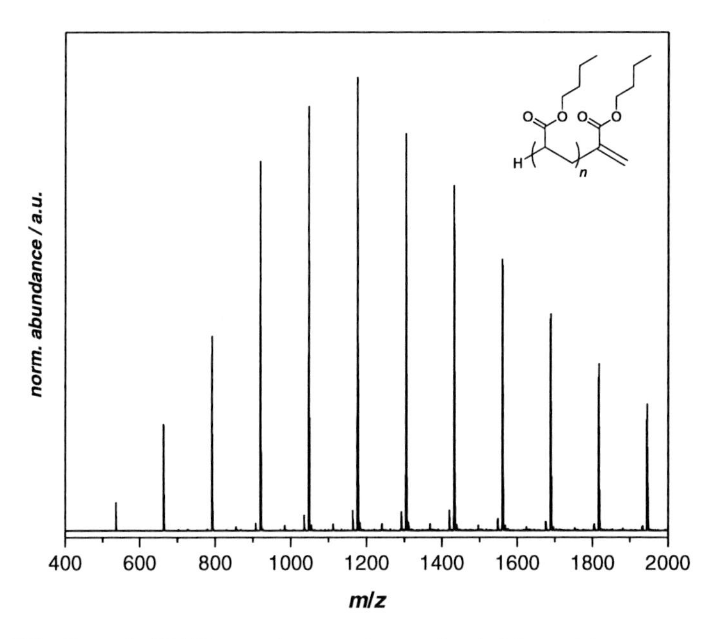

Figure 5-7. General ESI-MS overview spectrum of *n*-butyl acrylate macromonomer synthesized to full conversion in solution of butyl acetate with 5 wt% BA at 140 °C in an oxygen free atmosphere. The spectrum is shown in the *m/z* range of 400–2000.

For *m/z* assignments it is commonly sufficient to undertake a detailed analysis of one specific repeat unit of the mass spectrum. Predominantly, mass spectra recorded in the present work were analyzed in this manner with prior calibration in the mass range *m/z* = 200–2000, with exceptions if necessary. In addition to singly charged species, multiple charging is observed, commonly more relevant for higher molecular weight polymers. From the detected *m/z* values the according molecular weight *M* for each chain length can be calculated by the application of equation (5-2).

$$\frac{m}{z} = \frac{M + n \cdot M_a}{n} \tag{5-2}$$

The multiplicity *n* of the attached ion species is taken into account by the exact mass of M_a, i.e., Na^+ with $M_a = 22.989$. The accuracy of the experimental *m/z* is characterized by $\Delta\, m/z$ calculated from the difference of the theoretical and measured *m/z*s. For a good agreement of the assignments the $\Delta\, m/z$ needs to be low ($\Delta\, m/z < 0.3$).

The performance of the mass spectrometer and thus the appearance of the signals at a specific *m/z* value is influenced by the resolution of the applied mass analyzer. A medium resolution mass analyzer – as the linear quadrupole ion trap used herein – resolves the individual isotopes of the quasimolecular ion, whereas an unresolved peak at the average *m/z* value is detected for a poor resolution mass analyzer. Peak appearance at a different resolution is depicted in Figure 5-8.

The solvents used for ESI mass spectrometry need to be carefully chosen since the ionization process entails the evaporation of solvent forming quasimolecular ions. In the present work, the mass spectra were recorded from samples dissolved in a solvent mixture of tetrahydrofuran and methanol 3/2 (v/v) as convenient volatile organic solvents. The choice of solvents significantly affects the analyte charge state, signal intensity and stability of the gaseous ions.[198-199]

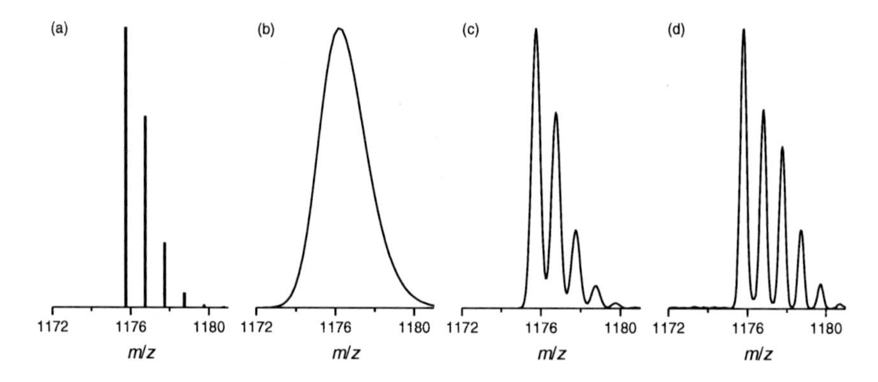

Figure 5-8. Peak appearance depending on the resolution of a mass analyzer visualized on the signal for the **MM^H** species $[C_{63}H_{108}O_{18}+Na]^+$. The isotopic pattern (a) in comparison to a Gaussian profile with a resolution of 1 Da (b) and 0.5 Da (c) and the experimental resolution obtained from a linear quadrupole ion trap (d).

The interpretation of SEC/ESI-MS spectra occurs in a similar manner. The measured RI signal represents the basis for the mass spectrometric analysis. The mass spectra can be generated through the integration of a specific elution volume and retention time, respectively. The obtained mass spectra related to a specific portion of the analyte are treated as described above.

6

Synthesis of a Macromonomer Library from High-Temperature Polymerization[⊥]

6.1 Introduction

It is a continuous aim in polymer chemistry to improve existing synthetic techniques or to establish new methods to produce well-defined polymers with predetermined molecular weights and high endgroup fidelity. To date, almost any polymer structure that can be thought of in terms of monomer type, chain length, dispersity and topology can be synthesized – at least in principle. However, with increasing complexity of the targeted structures, the synthetic effort that needs to be invested increases significantly, thus effectively limiting the type of polymers that can be generated on a laboratory or industrial scale.

Controlled/living radical polymerization methodologies, such as reversible addition fragmentation transfer (RAFT) polymerization,[3,113] atom transfer radical polymerization (ATRP)[2] or nitroxide-mediated polymerization (NMP)[1] to name the most common representatives, have given rise to a large number of variable polymers with highly defined structures. These synthetic protocols are frequently employed not only for their reliability to generate close-to-monodisperse polymers, but increasingly so, because of the high

[⊥] The work presented in this chapter was partly performed during my diploma thesis. Parts of this chapter were reproduced with permission from John Wiley & Sons.[11]

definition of endgroups that is usually obtained, allowing the polymers to be used as macromolecular building blocks in modular polymer conjugation reactions.[200]

While living/controlled polymerization techniques are almost exclusively used to generate polymers with homogeneous endgroups, other methods employing conventional radical polymerizations exist by which similar levels of endgroup fidelity can be achieved. Chiefari *et al.* introduced a method to create acrylic macromonomers in high yields by conventional polymerization at high temperatures.[10,70] A simplified reaction scheme of the macromonomer formation is depicted in Scheme 6-1.

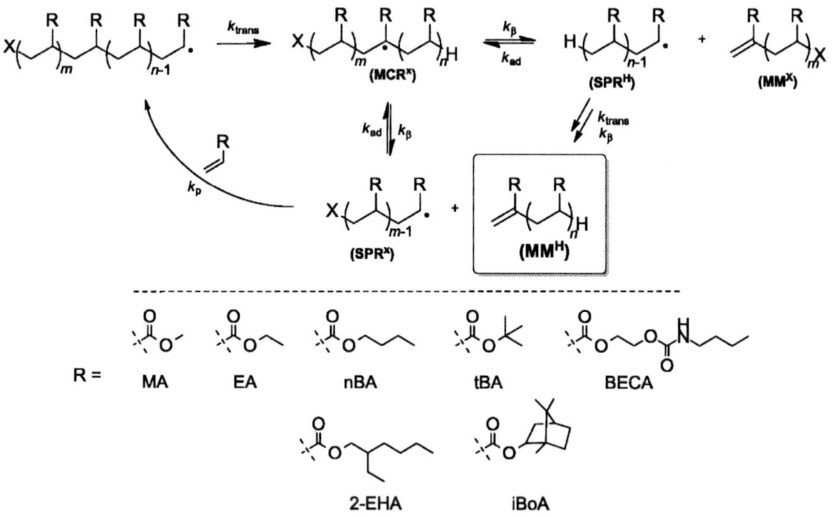

Scheme 6-1. Basic mechanism for the formation of $\mathbf{MM^H}$ in acrylate polymerizations alongside the monomers employed for the macromonomer synthesis in the present study.

In this chapter, it is demonstrated that the same synthetic approach as previously exclusively used for *n*-butyl acrylate[67] can be used for a large number of acrylic monomers and that successful macromonomer formation can be achieved for all esters given in the lower part of Scheme 6-1. Various macromonomers were synthesized via this route based on methyl (MA), ethyl (EA), *n*-butyl (BA), *t*-butyl (*t*BA), 2-ethylhexyl (2-EHA), isobornyl (iBoA) and 2-[[(butylamino)carbonyl]oxy]ethyl (BECA) acrylate. One important synthetic aspect that will be addressed is how the overall polymerization rate of the polymerization can be controlled, a situation that has not been satisfactorily addressed so far as the self-initiated polymerization often resulted in unpredictable reaction and inhibition times. Moreover, the reactivity of the macromonomers and their potential to be chain extended will be discussed.

6.2 Synthesis

Macromonomer Synthesis

For macromonomer synthesis the monomers were polymerized with 5×10^{-3} mol·L^{-1} AIBN in a 5 wt% solution of hexyl acetate (14.75 g, 98.8 mmol) (freed from oxygen by purging with argon for about 40 minutes prior to the reaction) in a pressure tube at 140 °C. The monomer (0.75 g, 5 wt%) was degassed by purging with argon for about 20 minutes in a separate vial sealed airtight with a septum at ambient temperature. The initiator AIBN (5×10^{-3} mol·L^{-1}) was dissolved in hexyl acetate and added to the reaction mixture at 140 °C. After 5 minutes the degassed monomer was added to the solvent and the pressure tube sealed airtight. The reaction solution was then stirred for about 16 hours. After the reaction, the solvent was removed in a vacuum oven at 45 °C. The purity of the synthesized macromonomer was determined by electrospray ionization mass spectrometry (ESI-MS).

Copolymer Synthesis

For the BA-EA copolymer macromonomer synthesis the BA macromonomer was provided in a 2.5 wt% solution of hexyl acetate (14.75 g, 98.8 mmol) (freed from oxygen by purging with argon for about 40 minutes prior to the reaction) in a pressure tube at 140 °C. The monomer ethyl acrylate (0.36 g, 2.5 wt%) was degassed by purging with argon for about 20 minutes in a separate vial sealed airtight with a septum at ambient temperature. The initiator AIBN (5×10^{-3} mol·L^{-1}) was dissolved in hexyl acetate and added to the reaction mixture at 140 °C. After 5 minutes the degassed monomer was added to the solvent and the pressure tube sealed airtight. The reaction solution was then typically stirred for about 16 hours. Product isolation and characterization was performed analogue to the (homo)macromonomer synthesis.

6.3 Results and Discussion

Based on the synthesis of *n*-butyl acrylate macromonomers a library of macromonomers was synthesized via high-temperature acrylate polymerization. The monomers that have been successfully converted into macromonomers are shown in Scheme 6-1.

During the polymerization several products may be formed.[67] Generally, a distinction is made between unsaturated and saturated products. The terminal double bonds of the oligomers are capable of being used in further transformations to generate polymeric architectures. In addition to the unsaturated species $\mathbf{MM}^{\mathbf{H}}$, which represents the main product of the synthesis, the species $\mathbf{MM}^{\mathbf{Hex}}$ can also be formed. This species carries a hexyl side chain on the terminal ester moiety, which results from a transfer to solvent reaction and thus initiation of the polymerization by a hexyl acetate radical fragment. In principle, when an acetate is used as solvent that carries the same ester side chain as the acrylate monomer, no transfer peak may be detected as the transfer-to-solvent product and $\mathbf{MM^{H}}$ are then chemically identical.

Thus, using hexyl acetate allows for detection of the amount of transfer to solvent taking place without disallowing the use of the product as a building block, as $\mathbf{MM^{H}}$ and $\mathbf{MM^{Hex}}$ will be of the same reactivity. More generally, $\mathbf{MM^{Hex}}$ is given as $\mathbf{MM^{X}}$, whereby X stands for any potential initiating radical. The formation of a saturated product is also possible, however, the saturated species $_{\mathbf{sat}}\mathbf{P}$ and $_{\mathbf{sat}}\mathbf{P^{Hex}}$ are only formed in small quantities. The chemical structures of all polymer species formed during the synthesis process are shown in Figure 6-1.

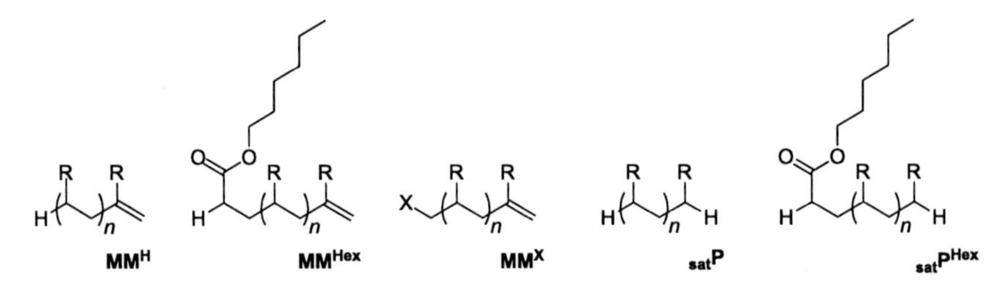

Figure 6-1. Chemical structures for the possible species identified in the mass spectra.

Since the high-temperature synthesis of acrylate-type macromonomers proceeds via an auto-initiated process, a free radical initiator is not strictly required. Nevertheless AIBN was added prior to the polymerization to remove minimal impurities in the reaction mixture. More detailed information on the effect of the initiator on the macromonomer formation will be discussed below. It may, however, safely be assumed that AIBN does not initiate any polymerization activity as all azo-compound is decomposed when the monomer is added to the solution. The half life of AIBN at 140 °C amounts to 0.15 minutes whereas the half life period of VAZO 88 is only slightly higher ($\tau_{½}$ = 0.30 min).[201-202]

As mentioned above, within the current study various acrylate macromonomers have been synthesized via the high-temperature procedure to full conversion. Table 6-1 gives a summary of the successfully synthesized macromonomers with detailed information on the individual product distributions.[203]

Table 6-1. Macromonomers synthesized to full conversion via high-temperature acrylate polymerization in solution of hexyl acetate with 5 wt% monomer at 140 °C in an oxygen free atmosphere. The individual species distribution after the high-temperature polymerization process is derived via ESI-MS. All numbers, except M_n, are given as percentage of the full product distribution.

	$\mathbf{MM^H}$	$\mathbf{MM^{Hex}}$	**Σ (MM)**	$\mathbf{_{sat}P}$	$\mathbf{_{sat}P^{Hex}}$	M_n /g·mol^{-1}
methyl acrylate (MA)	66	23	**89**	7	4	830
ethyl acrylate (EA)	81	9	**90**	9	1	1260
***n*-butyl acrylate (BA)**	88	5	**93**	7	0	1700
***t*-butyl acrylate (*t*BA)**	81	8	**90**	8	3	1790
2-ethylhexyl acrylate (EHA)	74	15	**89**	8	3	1450
isobornyl acrylate (iBoA)	53	29	**82**	11	7	1410
2-[[(butylamino)carbonyl] oxy]ethyl acrylate (BECA)	82	13	**95**	3	2	2100

The molecular weight of the samples is well within the range of the calibrated *m/z* range of the ESI-MS experimental set-up. Thus, ESI-MS provides a convenient avenue to analyze the macromonomers with respect to their endgroup purity.[68,72] The percentage of the 4 products – $\mathbf{MM^H}$, $\mathbf{MM^{Hex}}$, $\mathbf{_{sat}P}$ and $\mathbf{_{sat}P^{Hex}}$ – is estimated by averaging over the mass range of 3 repeating monomer units of the ESI-MS spectrum. Within each unit, the signal intensity was normalized and the content of each product calculated via the height of its corresponding peak.[71]

As a general tendency, the synthetic procedure delivers in all cases the unsaturated $\mathbf{MM^H}$ as a main product with the transfer product $\mathbf{MM^{Hex}}$ as the most pronounced side product. The best ratio of $\mathbf{MM^H}$: $\mathbf{MM^{Hex}}$ can be achieved with the well known *n*-butyl acrylate[67] and the ethyl acrylate system.

The most abundant peaks of the spectrum are represented to be single charged species of $\mathbf{MM^H}$ (88 %). A typical mass spectrum of *n*-butyl acrylate macromonomer in the *m/z* range from 400–2000 is shown in Figure 6-2 as a representative for all synthesized macromonomers.

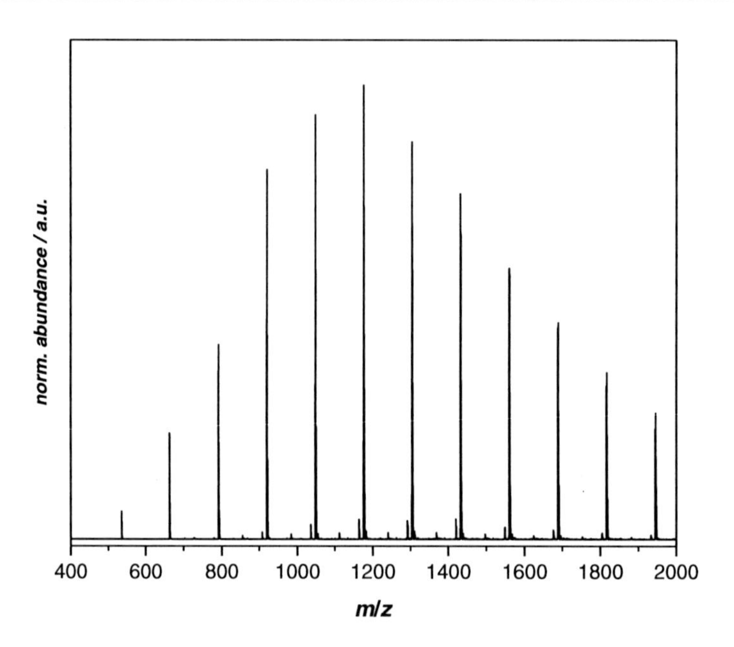

Figure 6-2. ESI-MS spectrum of BAMM synthesized to full conversion in solution of hexyl acetate with 5 wt% BA at 140 °C in an oxygen free atmosphere.

For a detailed inspection of the spectrum shown in Figure 6-2, a zoom spectrum of a repeat unit of the BA macromonomer either synthesized in butyl acetate (top) or hexyl acetate (bottom) is depicted in Figure 6-3.

The difference in the reaction solvent can clearly be seen in the mass spectra since the species $\mathbf{MM^{Hex}}$ referred to a transfer-to-solvent product are detectable in spectrum (b) of Figure 6-3 at m/z = 1203.8, whereas spectrum (a) of Figure 6-3 shows no appearance of this product species since the solvent – butyl acetate – carries the same ester side chain as the acrylate monomer; the transfer-to-solvent product and the $\mathbf{MM^{H}}$ are thus chemically identical. The assignments of the identified species in the mass spectra are collated in Table 6-2.

It should be noted that both $\mathbf{MM^{H}}$ and $\mathbf{MM^{Hex}}$ carry the specific terminal double bond, thus they can equally be used as potential synthetic building blocks; the Hex endgroup only differs in the terminal ester-side chain. All unsaturated side products are formed in small amounts.

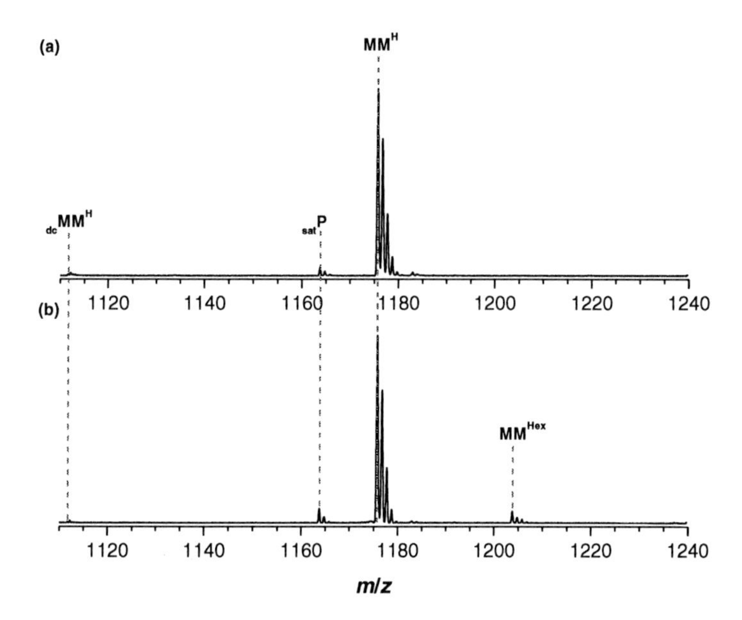

Figure 6-3. ESI-MS spectrum of BAMM synthesized to full conversion in solution of butyl acetate (a) or hexyl acetate (b) with 5 wt% BA at 140 °C in an oxygen free atmosphere. The exact *m/z* ratios associated with each species are collated in Table 6-2.

Table 6-2. Theoretical and experimental *m/z* ratios of the species of the BAMM found on the ESI-MS measurements.

	$m/z_{\text{theo.}}$	$m/z_{\text{exp.}}$	$\Delta\, m/z$
$[MM^{H}+2Na]^{2+}$	1111.70	1111.6	0.1
$[_{sat}P+Na]^{+}$	1163.74	1163.8	0.1
$[MM^{H}+Na]^{+}$	1175.74	1175.8	0.1
$[MM^{H}+2Na]^{2+}$	1175.74	1175.8	0.1
$[MM^{Hex}+Na]^{+}$	1203.77	1203.8	< 0.1

The characterization of the obtained macromonomers via ESI-MS clearly evidences a successful synthesis process by the appearance of the desired species $\mathbf{MM^{H}}$ as the most abundant signal. The unsaturated character of macromonomers can additionally be visualized by ^{1}H-NMR spectroscopy. The olefinc terminus is clearly visible at ≈ 5.62 ppm and ≈ 6.16 ppm representing the resonances of the protons on the geminal olefin. The resonances 2 to 5 are related to the CH_2 and CH_3, respectively, of the *n*-butyl ester side chain. A typical NMR spectrum of *n*-butyl acrylate is depicted in Figure 6-4 (the assignments of the detected resonances are embedded in the structural formula).

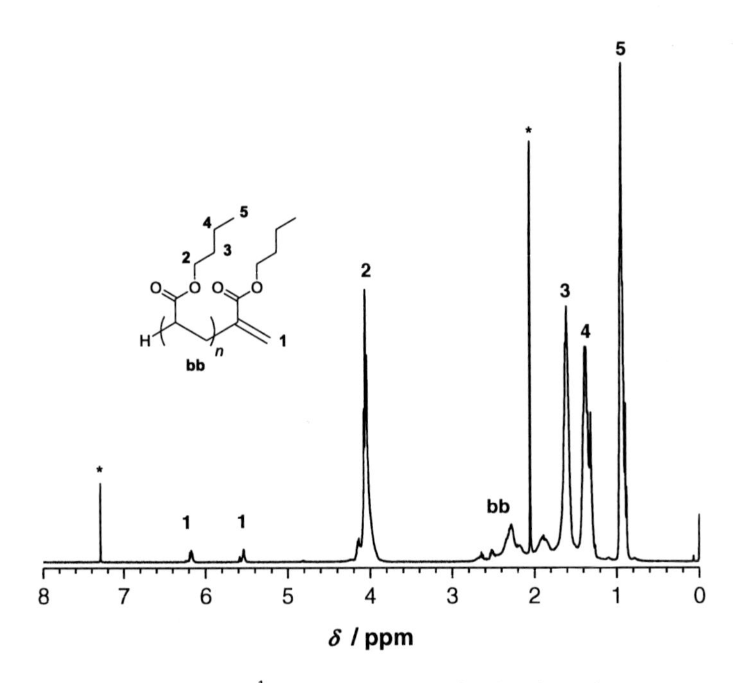

Figure 6-4. Typical 400 MHz ^{1}H-NMR spectrum of *n*-butyl acrylate macromonomer synthesized to full conversion in solution of butyl acetate with 5 wt% BA at 140 °C in an oxygen free atmosphere. The asterisks represent chloroform and water in chloroform, respectively. The structural formula with the detected resonances is inserted.

The molecular weight of the macromonomers is located between 800 g·mol^{-1} and 2000 g·mol^{-1} with a polydispersity of close to 1.6. A typical SEC elugram of *n*-butyl acrylate macromonomers synthesized via the high-temperature polymerization technique is shown in Figure 6-5 by taking the example of *n*-butyl acrylate macromonomer. Detailed ESI-MS characterization of all synthesized macromonomers can be found in ref. 11.

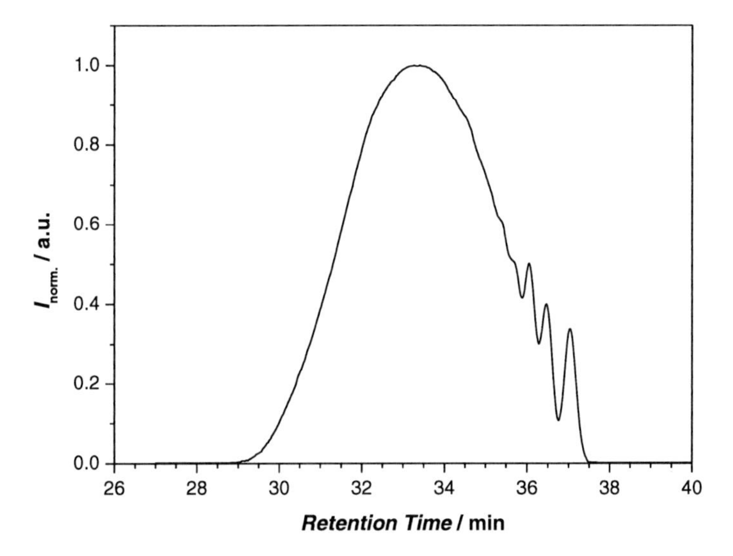

Figure 6-5. SEC elugram of the BAMM synthesized via the high-temperature acrylate polymerization at 140 °C in hexyl acetate with 5×10^{-3} mol·L^{-1} AIBN.

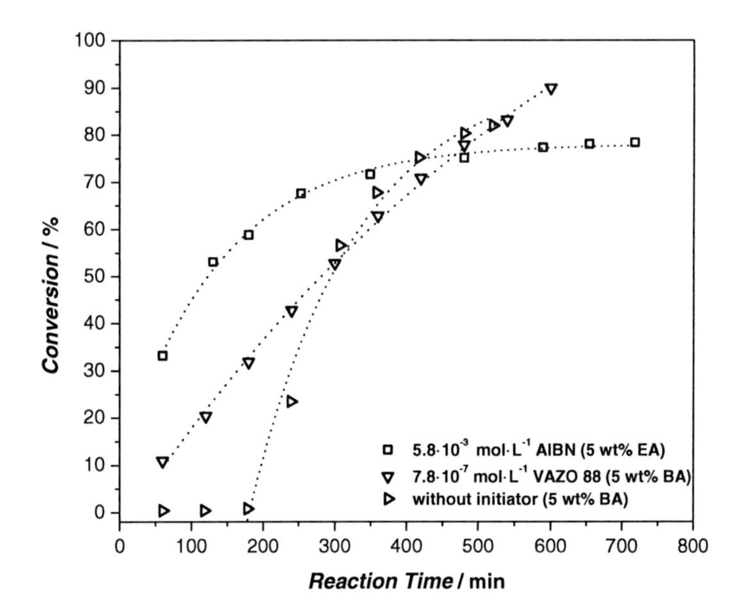

Figure 6-6. Conversion vs. time plot demonstrating the influence of the radical initiator on the conversion. The data were obtained via gravimetric measurements.

The product spectrum of all macromonomers are well in-line with the expectations based on Scheme 6-1.[64] Even though auto-initiation provides a feasible way of synthesizing **MM^{H}** in high purity (actually more uniform in terms of the saturated endgroups than when an initiator is employed), for the synthesis of macromonomer in high purity the use of radical initiator is beneficial, although it does not function as a starting point for macromolecular growth. In the present study, azoinitiators were used with the purpose of 'cleaning' the reaction mixture prior to the auto-initiated reaction. This effect is exemplified by the kinetic data given in Figure 6-6, which displays conversion time profiles for different macromonomer synthesis reactions determined by gravimetry.

Inspection of Figure 6-6 indicates that the radical initiator influences the starting point of the reaction. Three reactions are depicted that were carried out with two monomers (ethyl and *n*-butyl acrylate). Circles and triangles represent the synthesis of *n*-butyl acrylate macromonomer in a 5 wt% solution of hexyl acetate in an oxygen free atmosphere with and without radical initiator (VAZO 88 in this case). The closed squares depict the synthesis of ethyl acrylate macromonomer in a 5 wt% solution of hexyl acetate in an oxygen free atmosphere with comparatively high amounts of AIBN (5.8×10^{-3} mol·L^{-1}) as a radical purger. The synthesis without radical initiator displays a pronounced inhibition time that amounts to about 180 minutes before the start of the reaction. In comparison to the initiator-free reaction the synthesis in presence of 7.8×10^{-7} mol·L^{-1} VAZO 88 shows no inhibition time. Thus, the initiator leads to a faster start of polymerization and shortens the reaction time itself, an observation that supports the hypothesis of the azo-compound acting as a remover of impurities that would otherwise inhibit the reaction at the beginning. The conversion itself reaches close to 90 % in each reaction and is independent of the use of a radical source. In case of EA polymerization, the final conversion is somewhat lower. It should be noted that this is mostly due to a systematic effect of the gravimetry and that true conversions are most likely higher than given. From backbiting, a significant amount of acrylate dimers are formed which, in case of EA are partially volatile thus leading to a lowered apparent conversion after removal of the solvent *in vacuo*.

Since the initiator only acts as a purger and does not start the reaction itself, it is not overly important which and what amount of azo-compound is used in the synthesis as is evident from the conversion vs. time profile of the EA macromonomer reaction where comparatively high amounts of AIBN (5.8×10^{-3} mol·L^{-1}) are used (allowing for generation of a large number of primary radicals in a short period of time at 140 °C). Similar to the VAZO-88 supported polymerization of BA, a direct start of ethyl acrylate macromonomer formation is observed. If no initiator is added prior to EA polymerization, similar inhibition periods as with *n*-butyl acrylate would be observed.

As already mentioned, the generated macromonomers have potential use as convenient macromolecular building blocks in the generation of a wide variety of functionalized materials (for example surface grafting) and more complex macromolecular architectures.

To demonstrate the reactivity of the unsaturated oligomers in further reactions the *n*-butyl acrylate macromonomer was copolymerized with ethyl acrylate. The synthesis provides a pBA-*co*-pEA copolymer whose product spectrum was recorded by ESI-MS. Such a mass spectrum is depicted in Figure 6-7. Table 6-3 contains the *m*/*z* ratios of all peaks and the theoretical mass of the assigned structures.

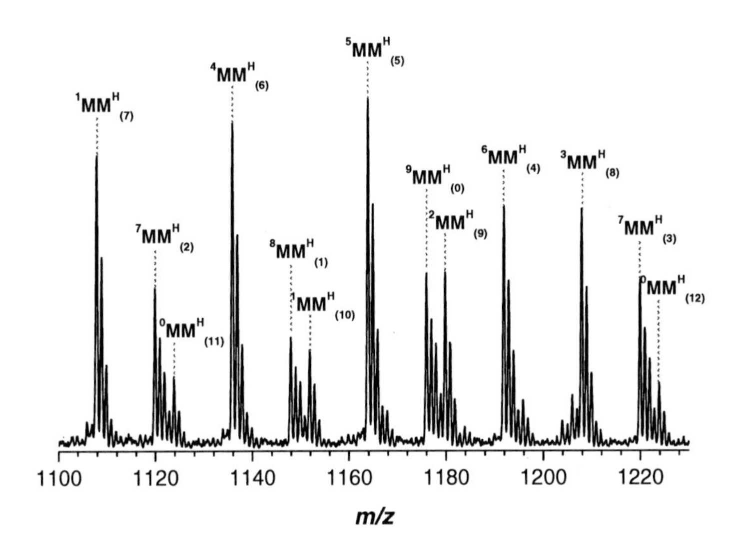

Figure 6-7. ESI-MS spectrum of pBA-*co*-pEA macromonomer synthesized with 2.5 wt% BA macromonomer and 2.5 wt% EA at 140 °C. Shown is a repeat unit of BA (M_{BA} = 128.08 Da). The subscript *m* in brackets indicates the number of EA repeating units in the copolymer. The subscript *n* up front denotes the number of BA units.

Table 6-3. Theoretical and experimental *m/z* ratios of the species of the pBA-*co*-pEA macromonomer determined via ESI-MS measurements. The resolution is close to 0.1 amu. *m* corresponds to the number of EA and *n* repeat units. For the chemical structures of the species refer to Figure 6-8.

	m/z **theo.**	***m/z*** **exp.**	**Δ *m/z***
$^{1}MM^{H}_{(7)}$ [Na]$^{+}$ m=7, n=1	1107.61	1107.8	0.2
$^{1}MM^{Hex}_{(8)}$ [Na]$^{+}$ m=8, n=1	1107.61	1107.8	0.2
$^{7}{}_{sat}P_{(2)}$ [Na]$^{+}$ m=2, n=7	1107.68	1107.8	0.1
$^{7}MM^{H}_{(2)}$ [Na]$^{+}$ m=2, n=7	1119.68	1119.8	0.1
$^{5}MM^{Hex}_{(3)}$ [Na]$^{+}$ m=3, n=5	1119.68	1119.8	0.1
$^{0}MM^{H}_{(11)}$ [Na]$^{+}$ m=11, n=0	1123.57	1123.8	0.2
$^{4}{}_{sat}P_{(6)}$ [Na]$^{+}$ m=6, n=4	1123.64	1123.8	0.2
$^{4}MM^{H}_{(6)}$ [Na]$^{+}$ m=6, n=4	1135.64	1135.8	0.2
$^{2}MM^{Hex}_{(7)}$ [Na]$^{+}$ m=7, n=2	1135.64	1135.8	0.2
$^{8}{}_{sat}P_{(1)}$ [Na]$^{+}$ m=1, n=8	1135.71	1135.8	0.1
$^{8}MM^{H}_{(1)}$ [Na]$^{+}$ m=1, n=8	1147.71	1147.9	0.1
$^{6}MM^{Hex}_{(2)}$ [Na]$^{+}$ m=2, n=6	1147.71	1147.9	0.1
$^{1}MM^{H}_{(10)}$ [Na]$^{+}$ m=10, n=1	1151.60	1151.8	0.2
$^{5}{}_{sat}P_{(5)}$ [Na]$^{+}$ m=5, n=5	1151.67	1151.8	0.1
$^{5}MM^{H}_{(5)}$ [Na]$^{+}$ m=5, n=5	1163.67	1163.9	0.2
$^{3}MM^{Hex}_{(6)}$ [Na]$^{+}$ m=6, n=3	1163.67	1163.9	0.2
$^{9}{}_{sat}P_{(0)}$ [Na]$^{+}$ m=0, n=9	1163.74	1163.9	0.2
$^{9}MM^{H}_{(0)}$ [Na]$^{+}$ m=0, n=9	1175.74	1176.0	0.3
$^{7}MM^{Hex}_{(1)}$ [Na]$^{+}$ m=1, n=7	1175.74	1176.0	0.3
$^{2}MM^{H}_{(9)}$ [Na]$^{+}$ m=9, n=2	1179.63	1179.8	0.2
$^{0}MM^{Hex}_{(10)}$ [Na]$^{+}$ m=10, n=0	1179.63	1179.8	0.2
$^{6}{}_{sat}P_{(4)}$ [Na]$^{+}$ m=4, n=6	1179.70	1179.8	0.1
$^{6}MM^{H}_{(4)}$ [Na]$^{+}$ m=4, n=6	1191.70	1191.9	0.2
$^{4}MM^{Hex}_{(5)}$ [Na]$^{+}$ m=5, n=4	1191.70	1191.9	0.2
$^{3}MM^{H}_{(8)}$ [Na]$^{+}$ m=8, n=3	1207.66	1207.9	0.2
$^{1}MM^{Hex}_{(9)}$ [Na]$^{+}$ m=9, n=1	1207.66	1207.9	0.2

	m/z theo.	m/z exp.	$\Delta m/z$
$^{7}{}_{sat}P_{(3)}$ $[Na]^+$ m=3, n=7	1207.73	1207.9	0.2
$^{7}MM^{H}{}_{(3)}$ $[Na]^+$ m=3, n=7	1219.73	1219.9	0.2
$^{5}MM^{Hex}{}_{(4)}$ $[Na]^+$ m=4, n=5	1219.73	1219.9	0.2
$^{0}MM^{H}{}_{(12)}$ $[Na]^+$ m=12, n=0	1223.60	1223.9	0.3
$^{4}{}_{sat}P_{(7)}$ $[Na]^+$ m=7, n=4	1223.69	1223.9	0.2

All product peaks in the spectrum can be assigned to macromonomer species (refer to Figure 6-8) and the large number of peaks per repeat unit is solely due to the presence of a copolymer mixture.

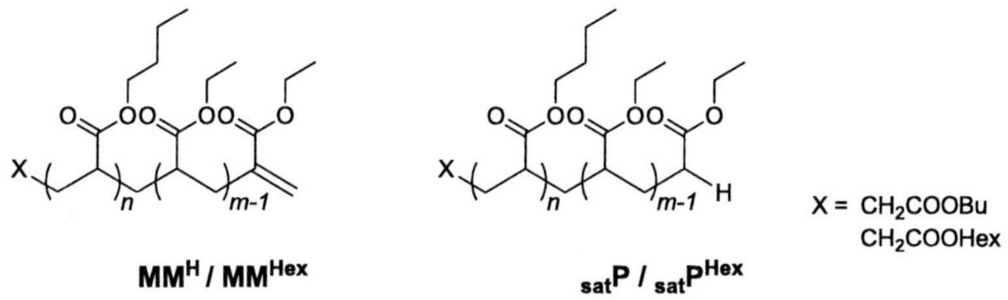

Figure 6-8. Chemical structures of the found products in the mass spectrum of pBA-*co*-pEA macromonomer.

For the peak assignments, each peak label gives the respective number of ethyl acrylate units in the backbone of the macromonomer (lower index on the right side, *m*) and the number of *n*-butyl acrylate units (upper index on the left side, *n*) respectively. Thus, **$^{0}MM^{H}{}_{(12)}$** represents pure EA macromonomer, whereas for example **$^{6}MM^{H}{}_{(4)}$** denotes a macromonomer that contains 4 EA units and the corresponding number of BA units (6 in case of the repeat unit starting from *m/z* 1100). The peak assignment is, however, more complicated. Due to isobaric overlap, each peak can also be assigned to a species with a terminal hexyl ester side chain. In addition, few (but not all) peaks overlap with **$_{sat}P$** species. Thus, a direct estimation of the amount of macromonomer relative to saturated products is not possible. However, the most abundant species, assigned to **$^{5}MM^{H}{}_{(5)}$**, does not overlap with any saturated copolymer structure, but only with **$^{9}{}_{sat}P_{(0)}$**, a species that is only found in minute quantities in the homomacromonomer spectrum. An additional indication for the high purity of macromonomer structures is given by analysis of the copolymer by ^{1}H-NMR, from which a degree of polymerization of 42 is calculated based on comparing the integrals of the vinylic protons with backbone peaks (refer to ref. 11).[10,204] This value is in good agreement with the observation of an apparent M_n of 5000 g·mol^{-1} from SEC.

The mass spectrometric analysis provides information on the composition of the copolymer, but it cannot resolve the sequence of monomers within the chains. At least partially, the original homomacromonomer can act as a degenerative transfer agent, in which case a blockpolymer structure would be obtained. Based on previous simulations,[64] it appears however to be more likely that the MCR that is formed upon addition of a (oligomeric) radical onto a macromonomer reacts with monomer preventing scission of the chain (which would result in degenerative transfer). In such a case, a classical random copolymerization between the initial monomer and the macromonomer chains is obtained. Further topological studies, even though not easily performed, appear to be of high importance to elucidate the exact structure of such copolymers in future work.

6.4 Conclusions

The auto-initiated high-temperature radical polymerization of acrylates allows for the synthesis of macromonomers end-functionalized with a geminal double bond. The synthesis requires a temperature of 140 °C and is carried out in a 5 wt% solution of hexyl actetate. Presently, a library of macromonomers was synthesized inter alia from methyl acrylate, ethyl acrylate, *n*-butyl acrylate, *t*-butyl acrylate, isobornyl acrylate, 2-ethylhexyl acrylate and 2-[[(butylamino)carbonyl]oxy]ethyl acrylate. The formation of the macromonomers follows an auto-initiation process and a radical initiator prevents the occurrence of an inhibition period and accelerates the start of reaction, but has no influence on the polymer product structure. The amount of macromonomers containing the geminal double bond lies in between 82 % and 95 %, depending on the monomer type. The achievable molecular weight of the macromonomers is located between 800 $g \cdot mol^{-1}$ and 2000 $g \cdot mol^{-1}$ with a polydispersity of close to 1.6.

Additionally, chain extension of macromonomers is shown to be a useful synthetic route as was realized in the copolymerization of *n*-butyl acrylate macromonomer with ethyl acrylate to form copolymeric macromonomer. The formation of the copolymer is not only useful for synthetic reasons, it also nicely demonstrates that the $\mathbf{MM^H}$ terminal bond is very reactive and is incorporated in growing polymer chains, a feature that is highly important when the $\mathbf{MM^H}$ is to be employed to build graft polymers or even more complex architectures.

7

High-Temperature Synthesis of Vinyl Terminated Polymers Based on Dendronized Acrylates

7.1 Introduction to Dendrimer Chemistry

7.1.1 Historical Background and Recent Developments

In the last decades, interest in dendrimer chemistry rapidly increased due to their inherent monodisperse constitution, high degree of branching and thermal and physical properties in comparison to linear polymers.[205-206]

Research in this specific area started to become known with the pioneering work of Vögtle. Vögtle and co-workers first reported these branched architectures in the late 1970s using the concept of repetitive growth which was then simply named 'cascade synthesis',[207] obtaining highly branched structures yet with low molecular weights. A few years later in the 1980s, further results in the area of dendrimers were published by Tomalia and colleagues at Dow Chemical.[208-209] Tomalia concentrated on poly(amidoamine) dendrimers (PAMAM) which were synthesized from either an ammonia or alternatively multi-functional amino core via divergent growth. At the same time, Newkome *et al.* reported the so-called 'arborol' systems based on methanetricarboxylate as the monomeric building block.[210-211] The obtained structures represented the first in detail investigated dendrimers. In contrast to the Tomalia-type dendrimers, Newkome applied convergent growth as a synthetic strategy. In

---

Parts of this chapter were reproduced by permission of the Royal Society of Chemistry.[12]

convergent growth, the periphery of the dendrimer is synthesized prior to a coupling step to a core. A detailed account of the synthetic methodologies will be provided later. The same synthetic pathway was used by Fréchet and co-workers in 1990,[212-213] in which they utilized 3,5-dihydroxybenzyl alcohol as monomer which was reacted to create a generation five dendrimer.

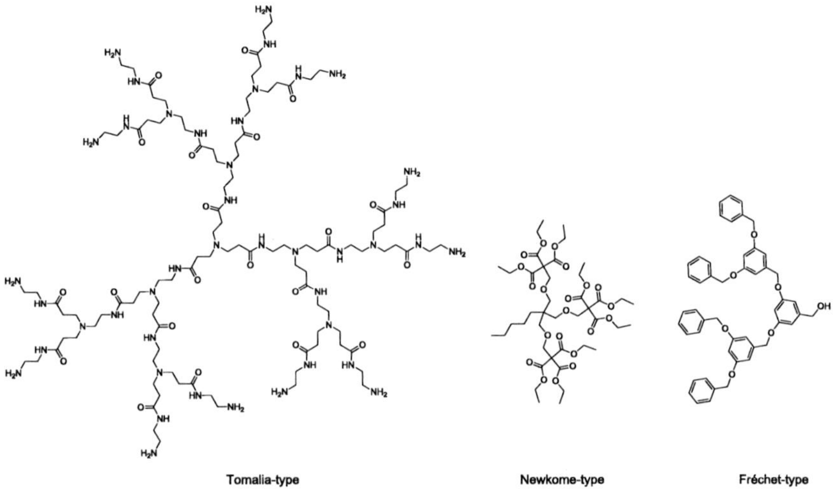

Figure 7-1. Chemical structure of a Tomalia-type poly(amidoamine) (PAMAM) dendrimer, Newkome-type 'arborol' dendrimer and Fréchet-type poly(aryl ether) denrimer (all structures represent 2nd generation).

7.1.2 Synthetic Pathways to Dendritic Structures

Dendritic macromolecules are globular in shape with high symmetry. The perfect structure features high endgroup fidelity on the periphery of the spherical structure.[214] The number and chemical nature of the endgroups in the molecule dictates the properties of the material. Depending on their structural conformation dendritic structures behave contrary to their linear counterparts. Dendritic structures can be divided into several substructures such as dendrimers, hyperbranched polymers, dendrigrafts and dendritic linear hybrids including dendronized polymers[215] which are depicted in Figure 7-2.

Each subclass features different chemical and physical properties and can be synthesized through different methodologies. The characteristics and synthetic strategies for the dendritic subclasses are described in the following.

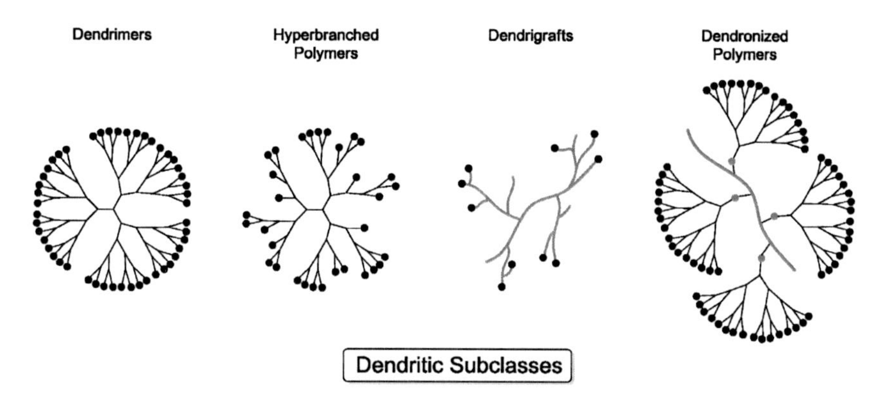

Figure 7-2. Substructures of the dendritic family.

7.1.2.1 Dendrimers and Dendrons

Dendrimers are perfectly symmetrical macromolecules with branches in the globular shape. Branched repeat units increase radially in layers from a core molecule towards the outside. The periphery is decorated with protected or deprotected endgroups.[216-220] Depending on the amount of radial branches the dendrimer is categorized into generations. Each generation emerges from a synthetic procedure containing activation/deactivation protocols. During the synthesis a high degree of perfection is obtained which implicates a nearly monodisperse macromolecule with a predictable molecular weight. The core molecule features functional groups which represent the linkage between the core and the branched arms, which are termed dendrons as a structural part of the dendrimer. A focal point on the dendron enables the formation of dendrimers by reaction with a core molecule. As shown in Scheme 7-1, the synthesis of dendrimers and dendrons can either follow a divergent (a) or a convergent (b) growth. Both synthetic strategies include repetitive stepwise growth via protection/deprotection steps of the monomers. After each generation purification steps are required to remove residual starting material, undesired side products such as defect dendrimers and incompletely decorated molecules. Commonly, column chromatography with a solvent gradient is used for the separation.

In the divergent approach dendrimer growth emanates from a multifunctional core molecule with at least two functional groups (Scheme 7-1 (a)) which react with an AB_x monomer. On each functionality – which are commonly identical – a dendron with the required number of generations is grown. The residual functionality B of the monomer is often protected which requires an additional deprotection step prior to further coupling reactions. In a subsequent step the deprotected periphery is reacted with another load of protected AB_x monomer. The sequence is repeated until the targeted number of generations is achieved. However, for higher generation dendrimers this synthetic strategy becomes more challenging due to the multitude of functionalities on the periphery. With increasing generations more monomer is required to ensure completion of the reaction. Nevertheless, current commercially available dendrimers are synthesized via divergent growth.[215]

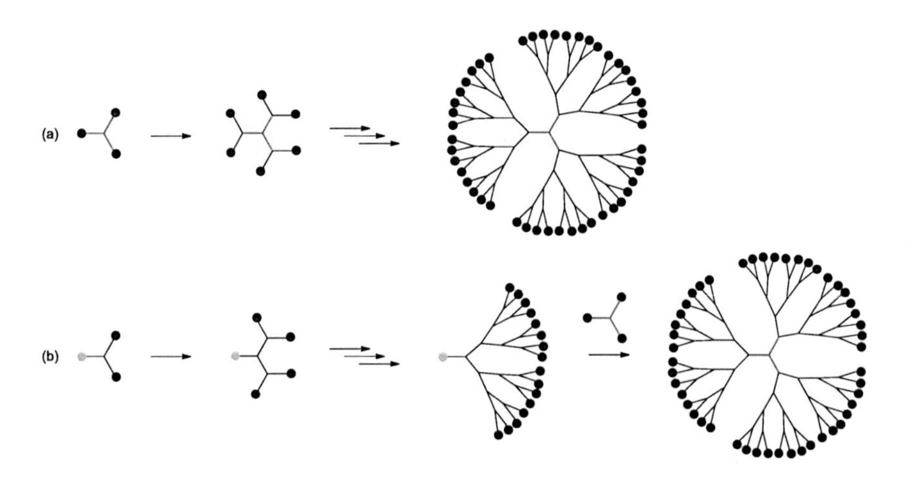

Scheme 7-1. Dendrimer syntheses via divergent growth (a) and convergent growth (b).

In contrast to the divergent method dendrons with a focal point are synthesized from the periphery for a convergent synthesis approach (Scheme 7-1 (b)).[212-213] The coupling to a multifunctional core molecule is realized in a subsequent step, first introduced by Hawker and Fréchet.[213] The dendron structures are synthesized through the same reaction steps as applied for the divergent growth, however purification after each generation growth is much more efficient. The convergent strategy benefits from the comparably large mass difference between the employed materials and the emerging dendrons. Nevertheless, convergent synthesis is limited to small generation dendrimers (up to 6 generations) due to sterical hindrance at the focal point. Coupling the dendrons to a core is achieved for six generations at the maximum.[205,208,215,221-223] Incorporation of a spacer between the dendron and its focal point could remedy this issue.

7.1.2.2 Hyperbranched and Dendrigraft Polymers

Hyperbranched polymers have a similar degree of branching as dendrimers and dendrons, yet with lower symmetry due to defects in branching.[224-231] Typically, hyperbranched polymers are synthesized via a straight forward one-pot procedure employing conventional polymerization methods of an AB_x monomer. Hyperbranched polymers feature a high branching density with potential branching in each repeat unit.[231] This subclass of dendritic structures shares similar properties with dendrimers and dendrons.[232] With increasing degree of branching they show more and more the same behavior as dendrimers. The high endgroup quantity enables better solubility compared to their linear counterparts. For further synthetic techniques the reader is referred to the literature.[233-234]

Dendrigraft polymers are well known as grafted polymers with a dendritic architecture.[235-236] Linear polymer chains feature linkages which can be employed for coupling branching elements. In the literature several other names appear for dendrigraft polymers, i.e., arborescent polymers, com-burst polymers, polymers with dendritic branching and hyper-Macs.[231] This subclass of the dendritic family is generally synthesized via the combination of living anionic/cationic polymerization techniques and grafting synthesis protocols.[236-237]

7.1.2.3 Dendronized Polymers

Dendronized polymers are defined as linear polymer chains with dendritic pendant unities attached on the polymer backbone.[238-239] The synthesis of polymers with higher generation dendrons is challenging due to sterical demand as already mentioned.[240-241] Three main strategies to achieve such structures are described in the literature. As shown in Scheme 7-2 these strategies include (a) the "grafting-to" approach – where pre-made dendrimers are convergently coupled via modular ligation methods to functionalized linear polymers, (b) the "grafting-from" approach – where the polymer backbone is stepwise functionalized via divergent growth or (c) the macromonomer approach where dendrons bearing a polymerizable functionality on the focal point are polymerized.[238-239,242] Purification of dendronized polymers can be carried out by a simple precipitation step which supersedes column chromatography.

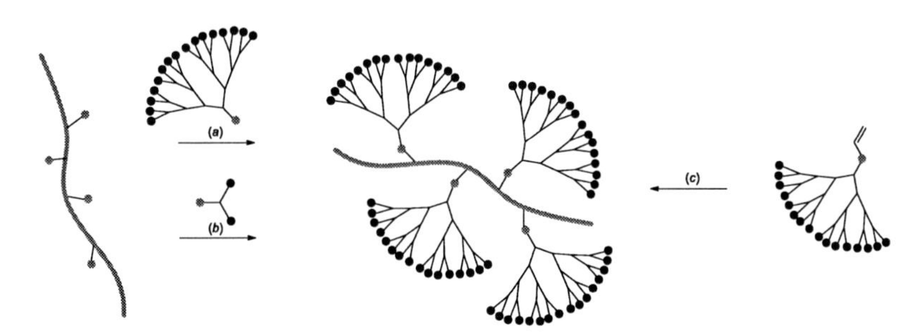

Scheme 7-2. Basic synthetic strategies for dendronized polymers via grafting-to (a), grafting-from (b) and macromonomer approaches (c).

Of the three synthetic strategies to achieve dendronized polymers, the macromonomer approach using radical or controlled/living radical polymerization methods is the most efficient synthetic procedure due to the high level of control over the dendritic side group incorporation within the macromolecule.[240-241,243-250] Dendronized polymers and oligomers are addressed in the following.

7.1.3 *Click* Chemistry and Dendrimer Synthesis

The research interests in dendrimer chemistry in the last 2 decades were driven by the further development of more efficient and simple synthetic protocols. For a long time the formation of dendritic structures was based on repetitive stepwise growth with deprotection and activation steps. New synthetic protocols – or rather philosophies – in organic chemistry were introduced in 2001 by Sharpless. The use of fast and efficient organic synthesis for dendrimer formation improved and – more important – accelerated the generation growth. Sharpless and co-workers coined the term '*click* chemistry' as a topic of a variety of versatile ligation reactions.[251] Several characteristics are required for a *click*-type reaction, such as:

- modular and quantitative reaction
- stoichiometric amounts of easy accessible chemicals
- selectivity towards desired products with high yields
- stereospecificity
- mild reaction conditions
- high tolerance towards functional groups
- simple product isolation

The most common and widely studied '*click* reaction' to date is the copper(I) catalyzed Huisgen 1,3-dipolar cycloaddition. Herein, an alkyne reacts with an azide (1,3-dipol) to form a 1,2,3-triazole linkage (see Scheme 7-3).

$$R_1{-}C{\equiv}CH \quad + \quad R_2{-}N_3 \xrightarrow{Cu(I)} \text{N=N, N–}R_2\text{, }R_1$$

Scheme 7-3. The copper(I)-catalyzed Huisgen 1,3-dipolar cycloaddition between an alkyne and an azide functionality.

In 2004, Hawker and co-workers first developed a dendrimer synthetic strategy based on *click* chemistry.[252] The convergent synthesis pathway depicted in Scheme 7-4 yielded highly pure dendrimers.

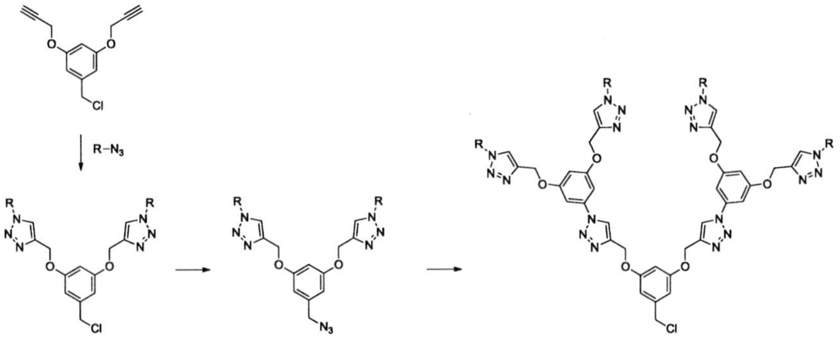

Scheme 7-4. Dendron synthesis via the use of *click* chemistry. The repetitive steps comprise a *click* reaction of an AB_2 monomer with an azide and a further deactivation of the B-functionality. Several other monomers were utilized in this study. The final dendron represents the 2nd generation.

The AB_2 monomer with acetylenes has been subjected to a copper-catalyzed azide-alkyne cycloaddition (CuAAC) with an azide in a first step. Further, the halogen has been substituted with an azide moiety activating the molecule for the next coupling step. Repetitive reaction steps were carried out up to 4 generations.

In 2005 a divergent synthetic strategy of Fréchet-type dendrimers based on *click* chemistry has been published by Wooley and co-workers.[253] The acetylene functionality A of the AB_2 monomer was coupled to a bifunctional azide core, with generations being formed from cycles of CuAAC and deactivation of the B functionality.

The CuAAC reaction can also be beneficial for accelerating more conventional dendrimer growth which has been described by Antoni and colleagues.[254] For a fast and efficient generation growth they combined traditional esterification/etherification steps in combination with *click* chemistry.

Additional to the conventional esterification procedure, thiol-ene *click* reactions have been employed for dendrimer formation.[255-256] By now, the most rapid and efficient dendrimer synthesis has been reported by Antoni and co-workers in 2010.[257] AB_2 and CD_2 monomers with orthogonally clickable groups were subjected to CuAAC and thiol-ene sequences accomplishing a 6th generation dendrimer within one day.

7.2 Dendronized Macromonomers via High-Temperature Polymerization of Dendronized Acrylates

In the following, a combination of dendrimer chemistry and high-temperature acrylate polymerization is shown to extend the existing macromonomer library.[11] The synthesis of dendronized acrylates based on 2,2-bis(hydroxymethyl)propionic monomer (bis-MPA) and their application in high-temperature acrylate polymerization yielding comb-macromonomers will be discussed. The combination of convergent acetylene bearing dendron synthesis, *click* chemistry[251,258] and high-temperature acrylate polymerization is a versatile technique to generate dendronized vinyl terminated polymers. Orthogonal modular ligation appeared to be a highly efficient synthetic pathway to generate the required dendritic acrylate monomers.[259-260] Scheme 7-5 depicts the general synthetic strategy followed in the current work.

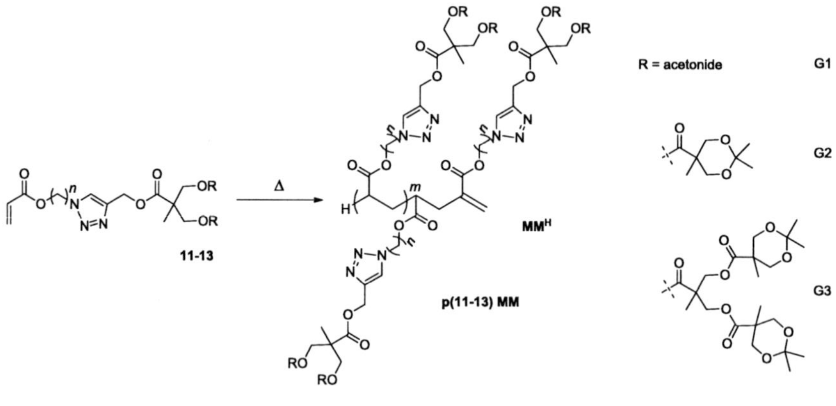

OR OR OR OR R = acetonide G1 G2 G3 11-13 Δ MM^H p(11-13) MM

Scheme 7-5. High-temperature acrylate polymerization of dendronized acrylates **11-13** (the numbering is consistent with the compound numbering during the synthesis (see Scheme 7-6)) forming vinyl terminated polymers/oligomers p(**11-13**) MM. Polymerization conditions: hexyl acetate, AIBN, 140 °C, 72 h.

Typically, the performance of a (free radical) polymerization in terms of monomer to polymer conversion decreases with higher generations of dendrons within the monomer due to steric hinderance.[240] To avoid this limiting factor a carbon spacer between the polymerizable endgroup and the dendritic structure of the macromonomer is a sought to be a tool to improve the polymerization properties.[240-241] In the current work a carbon 9 chain (9 $-CH_2-$ between the acrylic moiety and the dendritic structure) and a carbon 6 chain is employed, respectively, as a spacing unit. In the literature vinylic dendrons up to generation 4 have been reported to readily polymerize via convenient and controlled radical polymerizations, whereas higher generations are limited in their polymerization efficiency.[238]

Herein, dendronized acrylates were synthesized up to generation 3 and the subsequent macromonomer formation for all prepared generations was investigated. In addition, the copolymerization of generation 1 dendronized acrylates with ethyl acrylate was explored. In addition, the current study entails the in-depth assessment of the novel dendritic acrylate structures and – most importantly and as one of its core elements – a detailed structural product analysis of the prepared dendritic macromonomers via SEC as well as (SEC/)ESI-MS, allowing a quantification of the efficiency and viability of the macromonomer formation process.

7.2.1 Nomenclature

The nomenclature of the dendritic structures and dendronized acrylates is as follows: $(R)_m$ -[G*X*]-acetylene for acetylene terminated dendrons where R represents the external functional group, e.g., An for acetonide protection and OH for a hydroxyl function with *m* chain end functionalities. *X* indicates the generation of the dendritic structure. $(R)_m$-[G*X*]-triazole-alkyl acrylate (C_nG*X*) represents the corresponding dendronized acrylate with the above described nomenclature.

For the vinyl terminated polymers/oligomers the following nomenclature applies: the macromonomers $\mathbf{MM}^{H}_{nGX\text{-}mAn}$ and $\mathbf{MM}^{Hex}_{nGX\text{-}mAn}$ are terminated by a vinyl terminus and a proton or a hexyl side chain and initiated via a transfer to solvent reaction, respectively. Saturated products are labeled with $_{sat}\mathbf{P}_{nGX\text{-}mAn}$ and $_{sat}\mathbf{P}^{Hex}_{nGX\text{-}mAn}$, respectively, with the above mentioned restrictions according to the nomenclature for dendronized acrylates. The subscript of each species represents the amount and nature of the repeat units incorporated in the polymer chain. Thus, the general notation is: *n*(G*X*)-*m*(An) where *n* represents the respective overall number of repeat units G*X* lowered by *m* deprotected acetonide on the dendron surface for the homopolymer and *n*(G*X*)-*m*(An)+*EA* including the respective number of ethyl acrylate (EA) repeat units for the copolymer (see Figure 7-3 and Figure 7-13).

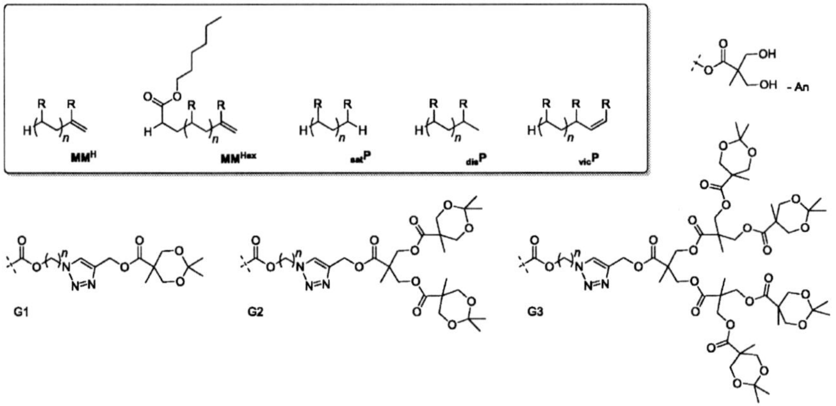

Figure 7-3. Chemical structures of the identified products in the mass spectrum of dendronized macromonomers.

7.2.2 Synthesis

The dendrons **5-7** (

Scheme **7-6**) have been synthesized according to literature.[261]

General Procedure for the Synthesis of Alkyl Azides by Nucleophilic Substitution

6-Azido-1-hexanol (9a). 6-Bromo-1-hexanol **8a** (10.0 g, 0.055 mol, 1 eq) was dissolved in DMSO (100 mL) and heated to 40 °C. Sodium azide (17.9 g, 0.276 mol, 5 eq) was added stepwise and the reaction mixture was stirred under reflux for 20 h. The organic phase was extracted with ether (3 × 250 mL), dried over $MgSO_4$ and concentrated under reduced pressure. **9a** was isolated as a colorless oil (82 %).[262]

^{1}H-NMR (400 MHz, $CDCl_3$): [δ, ppm] = 3.64 (t, J = 6.5 Hz, 2H, HO-CH_2-), 3.27 (t, J = 6.9 Hz, 2H, -CH_2-N_3), 1.64 - 1.54 (m, 4H, HO-CH_2-CH_2-, -CH_2-CH_2-N_3), 1.43 - 1.34 (m, 4H, -$(CH_2)_2$-CH_2-CH_2-N_3). ^{13}C-NMR (100 MHz, $CDCl_3$) δ 62.90 (1C, HO-CH_2-), 51.51 (1C, -CH_2-N_3), 32.68 (1C, -CH_2-CH_2-N_3), 28.94 (1C, -CH_2-CH_2-CH_2-N_3), 26.65 (1C, HO-CH_2-CH_2-CH_2), 25.46 (1C, HO-CH_2-CH_2-).

9-Azido-1-nonanol (9b) was prepared from 9-bromo-1-nonanol **8b** according to the *general procedure for the synthesis of alkyl azides by nucleophilic substitution* to give **9b** as a colorless oil (94 %).

^{1}H-NMR (400 MHz, $CDCl_3$): [δ, ppm] = 3.63 (t, J = 6.6 Hz, 2H, HO-CH_2-), 3.25 (t, J = 7.0 Hz, 2H, -CH_2-N_3), 1.63 - 1.52 (m, 4H, HO-CH_2-CH_2-, -CH_2-CH_2-N_3), 1.33 (d, J = 15.7 Hz, 10H, -$(CH_2)_5$-CH_2-CH_2-N_3). ^{13}C-NMR (100 MHz, $CDCl_3$) δ 63.20 (1C, HO-CH_2-), 51.62 (1C,

-*C*H$_2$-N$_3$), 32.91 (1C, HO-CH$_2$-*C*H$_2$-), 29.51 (1C, -*C*H$_2$-CH$_2$-N$_3$), 29.22 (2C, HO-(CH$_2$)$_3$-(*C*H$_2$)$_2$-), 28.97 (1C, -*C*H$_2$-(CH$_2$)$_3$-N$_3$), 26.84 (1C, -*C*H$_2$-(CH$_2$)$_2$-N$_3$), 25.84 (1C, HO-(CH$_2$)$_2$-*C*H$_2$-).

General Esterification Procedure for Acrylate Preparation

6-Azido-1-hexyl acrylate (10a). 6-Azido-1-hexanol **9a** (6.0 g, 0.042 mol, 1.0 eq), NEt$_3$ (6.4 g, 0.063 mol, 1.5 eq), DMAP (1.0 g, 0.008 mol, 0.2 eq) and hydroquinone (100 ppm) were dissolved in DCM (120 mL) and stirred at ambient temperature. Acryloyl chloride (5.7 g, 0.063 mol, 1.5 eq) was slowly added to the round bottom flask. The reaction mixture was stirred for 24 h. The solution was diluted with DCM (250 mL) and extracted with 10 % NaHSO$_4$ (3 × 200 mL), 10 % NaHCO$_3$ (3 × 200 mL) and brine (3 × 200 mL). The organic phase was dried over MgSO$_4$, filtered and concentrated *in vacuo*. The crude product was purified by column chromatography on silica gel with ethyl acetate/heptane 3/7 (v/v) to give **10a** as a colorless oil (81 %).

^{1}H-NMR (400 MHz, CDCl$_3$): [δ, ppm] = 6.40 (dd, J = 17.3, 1.4 Hz, 1H, *C*H$_2$=CH-), 6.12 (dd, J = 17.3, 10.4 Hz, 1H, CH$_2$=*C*H-), 5.82 (dd, J = 10.4, 1.5 Hz, 1H, *C*H$_2$=CH-), 4.16 (t, J = 6.6 Hz, 2H, -*C*H$_2$-OCO-), 3.27 (t, J = 6.9 Hz, 2H, -*C*H$_2$-N$_3$), 1.72 – 1.67 (m, 2H, -*C*H$_2$-CH$_2$-OCO-), 1.62 (dt, J = 10.9, 7.7 Hz, 2H, -*C*H$_2$-CH$_2$-N$_3$), 1.42 (dt, J = 7.2, 3.5 Hz, 4H, -(*C*H$_2$)$_2$-CH$_2$-CH$_2$-N$_3$).

9-Azido-1-nonanyl acrylate (10b) was prepared from 9-azido-1-nonanol **9b** according to the *general esterification procedure for acrylate preparation* to give **10b** as a colorless oil (80 %).

^{1}H-NMR (400 MHz, CDCl$_3$): [δ, ppm] = 6.39 (dd, J = 17.3, 1.4 Hz, 1H, *C*H$_2$=CH-), 6.12 (dd, J = 17.3, 10.4 Hz, 1H, CH$_2$=*C*H-), 5.81 (dd, J = 10.4, 1.4 Hz, 1H, *C*H$_2$=CH-), 4.14 (t, J = 6.7 Hz, 2H, -*C*H$_2$-OCO-), 3.25 (t, J = 6.9 Hz, 2H, -*C*H$_2$-N$_3$), 1.62 (ddd, J = 14.4, 12.7, 7.2 Hz, 4H, HO-CH$_2$-*C*H$_2$-, -*C*H$_2$-CH$_2$-N$_3$), 1.33 (d, J = 17.0 Hz, 10H, -(*C*H$_2$)$_5$-CH$_2$-CH$_2$-N$_3$).

General Procedure for the Copper-Catalyzed Azide-Alkyne Cycloaddition (CuAAC)

(An)$_1$-[G1]-Triazole-hexyl acrylate (11a). (An)$_1$-[G1]-acetylene **5** (3.0 g, 0.014 mol, 1.1 eq) and 6-azido-1-hexyl acrylate **10a** (2.5 g, 0.013 mol, 1.0 eq) were dissolved in 50 mL THF:H$_2$O (3:1). CuSO$_4$·5H$_2$O (0.3 g, 1.270 mmol, 0.1 eq) and sodium ascorbate (1.7 g, 6.340 mmol, 0.5 eq) were added to the reaction mixture and stirred over night at ambient temperature. The solvents were evaporated under reduced pressure and the crude product was purified by column chromatography on silica gel starting with ethyl acetate/heptane 2/8 (v/v) and gradually increasing polarity. The product was isolated as colorless oil (86 %).

^{1}H-NMR (400 MHz, CDCl$_3$): [δ, ppm] = 7.60 (s, 1H, -N-*C*H=C), 6.39 (dd, J = 17.3, 1.4 Hz, 1H, *C*H$_2$=CH-), 6.11 (dd, J = 17.3, 10.4 Hz, 1H, CH$_2$=*C*H-), 5.82 (dd, J = 10.4, 1.4 Hz, 1H, *C*H$_2$=CH-), 5.30 (s, 2H, -*C*H$_2$-OCO-), 4.34 (t, J = 7.2 Hz, 2H, -*C*H$_2$-N), 4.18 (d, J = 11.9 Hz, 2H, -C-*C*H$_2$-O- 1st generation), 4.13 (t, J = 6.6 Hz, 2H, O-*C*H$_2$-), 3.63 (d, J = 11.9 Hz, 2H, -C-*C*H$_2$-O- 1st generation), 1.96 - 1.87 (m, 2H, -*C*H$_2$-CH$_2$-N), 1.70 - 1.61 (m, 2H, O-CH$_2$-*C*H$_2$-), 1.38 (d, J = 27.6 Hz, 10H, C(*C*H$_3$)$_2$, -(*C*H$_2$)$_2$-), 1.13 (s, 3H, C-*C*H$_3$ 1st generation). ^{13}C-NMR (100 MHz,

$CDCl_3$): [*δ*, ppm] = 174.34 (1C, -CH_2O*C*O-), 166.40 (1C, H_2C=C-*C*O-), 143.07 (1C, N-CH=*C*), 130.83 (1C, H_2*C*=CH-), 128.60 (1C, H_2C=*C*H-), 123.59 (1C, N-*C*H=C), 98.21 (1C, *C*(CH_3)$_2$), 66.12 (2C, C-(CH_2O)$_2$), 64.37 (1C, -CH_2-OCO-), 58.35 (1C, -CH=C-*C*H_2-), 50.39 (1C, -*C*H_2N), 42.06 (1C, -OCO-*C*-), 30.27 (1C, -O-CH_2-*C*H_2), 28.52 (1C, -*C*H_2-CH_2-N), 26.26 (1C, -*C*H_2-(CH_2)$_2$-N), 25.53 (1C, C(*C*H_3)$_2$), 25.31 (1C, C(*C*H_3)$_2$), 22.24 (1C, -O-(CH_2)-*C*H_2-), 18.55 (1C, -OCO-C*C*H_3). ESI-MS – $[M+Na]^+$ theo. 432.21 Da, exp. 432.25 Da.

(An)$_1$-[G1]-Triazole-nonanyl acrylate (11b) was synthesized from (An)$_1$-[G1]-acetylene **5** and 9-azido-1-nonanyl acrylate **10b** according to the *general procedure for the Copper-Catalyzed Azide-Alkyne Cycloaddition (CuAAC)* to give **11b** as a colorless oil (75 %).

^{1}H-NMR (400 MHz, $CDCl_3$): [*δ*, ppm] = 7.60 (s, 1H, -N-C*H*=C), 6.39 (dd, *J* = 17.3, 1.4 Hz, 1H, C*H*$_2$=CH-), 6.11 (dd, *J* = 17.3, 10.4 Hz, 1H, CH_2=C*H*-), 5.81 (dd, *J* = 10.4, 1.4 Hz, 1H, C*H*$_2$=CH-), 5.30 (s, 2H, -C*H*$_2$-OCO-), 4.32 (t, *J* = 7.3 Hz, 2H, -C*H*$_2$-N), 4.18 (d, *J* = 11.8 Hz, 2H, C-C*H*$_2$-O- 1st generation), 4.13 (t, *J* = 6.7 Hz, 2H, -O-C*H*$_2$-), 3.63 (d, *J* = 11.8 Hz, 2H, -C-C*H*$_2$-O- 1st generation), 1.94 - 1.81 (m, 2H, -C*H*$_2$-CH_2-N), 1.64 (dd, *J* = 14.3, 6.9 Hz, 2H, -O-CH_2-C*H*$_2$-), 1.42 (s, 3H, C(C*H*$_3$)$_2$), 1.35 (s, 3H, C(C*H*$_3$)$_2$), 1.29 (d, *J* = 3.8 Hz, 10H, -(C*H*$_2$)$_5$-), 1.13 (s, 3H, C-C*H*$_3$ 1st generation). ^{13}C-NMR (100 MHz, $CDCl_3$): [*δ*, ppm] = 174.33, 166.49, 142.98, 130.66, 128.72, 123.59, 98.22, 66.11, 64.75, 58.34, 50.53, 42.05, 30.35, 29.28, 29.00, 28.68, 26.55, 25.97, 25.31, 22.23, 18.54. ESI-MS – $[M+Na]^+$ theo. 474.26 Da exp. 474.33 Da.

(An)$_2$-[G2]-Triazole-hexyl acrylate (12a) was synthesized from (An)$_2$-[G2]-acetylene **6** and 6-azido-1-hexyl acrylate **10a** according to the *general procedure for the Copper-Catalyzed Azide-Alkyne Cycloaddition (CuAAC)* to give **12a** as a colorless oil (75 %).

^{1}H-NMR (400 MHz, $CDCl_3$): [*δ*, ppm] = 7.62 (s, 1H, N-C*H*=C), 6.39 (dd, *J* = 17.3, 1.4 Hz, 1H, C*H*$_2$=CH-), 6.11 (dd, *J* = 17.3, 10.4 Hz, 1H, CH_2=C*H*-), 5.82 (dd, *J* = 10.4, 1.4 Hz, 1H, C*H*$_2$=CH-), 5.26 (s, 2H, -C*H*$_2$-OCO-), 4.34 (dd, *J* = 4.3, 2.9 Hz, 2H, -C*H*$_2$-N), 4.31 (s, 4H, -C-C*H*$_2$-O- 1st generation), 4.15 (d, *J* = 6.6 Hz, 2H, -O-C*H*$_2$-), 4.13 - 4.07 (m, 4H, -C-C*H*$_2$-O- 2nd generation), 3.58 (d, *J* = 12.2 Hz, 4H, -C-C*H*$_2$-O- 2nd generation), 1.98 - 1.87 (m, 2H, -C*H*$_2$-CH_2-N-), 1.74 - 1.62 (m, 2H, -O-CH_2-C*H*$_2$-), 1.41 (s, 6H, C(C*H*$_3$)$_2$ 2nd generation), 1.39 (s, 4H, -(C*H*$_2$)$_2$-), 1.34 (s, 6H, C(C*H*$_3$)$_2$ 2nd generation), 1.27 (s, 3H, C-C*H*$_3$ 1st generation), 1.10 (s, 6H, C-C*H*$_3$ 2nd generation). ^{13}C-NMR (100 MHz, $CDCl_3$): [*δ*, ppm] = 173.64, 166.42, 130.85, 128.60, 123.90, 116.29, 98.25, 66.07, 65.36, 64.39, 58.71, 50.41, 46.95, 42.16, 30.27, 28.54, 26.32, 25.49, 22.12, 18.61. ESI-MS – $[M+Na]^+$ theo. 704.34 Da exp. 704.42 Da.

(An)$_2$-[G2]-Triazole-nonanyl acrylate (12b) was synthesized from (An)$_2$-[G2]-acetylene **6** and 9-azido-1-nonanyl acrylate **10b** according to the *general procedure for the Copper-Catalyzed Azide-Alkyne Cycloaddition (CuAAC)* to give **12b** as a colorless oil (65 %).

^{1}H-NMR (400 MHz, $CDCl_3$): [*δ*, ppm] = 7.61 (s, 1H, N-C*H*=C), 6.39 (dd, *J* = 17.3, 1.4 Hz, 1H, C*H*$_2$=CH-), 6.12 (dd, *J* = 17.3, 10.4 Hz, 1H, CH_2=C*H*-), 5.82 (dd, *J* = 10.4, 1.4 Hz, 1H, C*H*$_2$=CH-), 5.26 (s, 2H, -C*H*$_2$-OCO-), 4.33 (d, *J* = 7.6 Hz, 2H, -C*H*$_2$-N), 4.31 (s, 4H, C-C*H*$_2$-O- 1st generation), 4.15 (d, *J* = 6.7 Hz, 2H, O-C*H*$_2$-), 4.12 - 4.07 (m, 4H, -C-C*H*$_2$-O- 2nd generation),

3.58 (d, *J* = 12.2 Hz, 4H, -C-CH_2-O- 2nd generation), 1.94 - 1.86 (m, 2H, -CH_2-CH_2-N), 1.70 - 1.62 (m, 2H, O-CH_2-CH_2-), 1.41 (s, 6H, C-CH_3 2nd generation), 1.34 (s, 16H, $C(CH_3)_2$ 2nd generation, -$(CH_2)_5$-), 1.27 (s, 3H, C-CH_3 1st generation), 1.10 (s, 6H, $C(CH_3)_2$ 2nd generation). ^{13}C-NMR (100 MHz, $CDCl_3$): [*δ*, ppm] = 173.63, 172.70, 142.43, 130.67, 128.72, 123.87, 98.24, 66.06, 65.36, 64.75, 58.72, 50.54, 46.94, 42.15, 30.37, 29.39, 29.04, 28.70, 26.62, 25.99, 25.42, 22.11, 18.60, 17.80. ESI-MS – $[M+Na]^+$ theo. 746.38 Da exp. 746.36 Da.

$(An)_4$-[G3]-Triazole-hexyl acrylate (13a) was synthesized from $(An)_4$-[G3]-acetylene **7** and 6-azido-1-hexyl acrylate **10a** according to the *general procedure for the Copper-Catalyzed Azide-Alkyne Cycloaddition (CuAAC)* to give **13a** as a colorless oil (50 %).

^{1}H-NMR (400 MHz, $CDCl_3$): [*δ*, ppm] = 7.72 (s, 1H, -N-*CH*=C), 6.39 (dd, *J* = 17.3, 1.4 Hz, 1H, CH_2=CH-), 6.11 (dd, *J* = 17.3, 10.4 Hz, 1H, CH_2=*CH*-), 5.82 (dd, *J* = 10.4, 1.4 Hz, 1H, CH_2=CH-), 5.26 (s, 2H, CH_2-OCO), 4.37 (t, *J* = 7.3 Hz, 2H, -CH_2-N), 4.29 - 4.24 (m, 10H, C-CH_2-O- 2nd generation, O-CH_2-), 4.23 (d, *J* = 4.0 Hz, 2H, C-CH_2-O- 1st generation), 4.17 - 4.11 (m, 10H, C-CH_2-O- 3nd generation, 1st generation), 3.62 (d, *J* = 12.2 Hz, 8H, C-CH_2-O- 3nd generation), 1.98 - 1.87 (m, 2H, -CH_2-CH_2-N), 1.71 - 1.62 (m, 6H, O-CH_2-$(CH_2)_3$-), 1.41 (s, 12H, $C(CH_3)_2$ 3rd generation), 1.34 (s, 12H, $C(CH_3)_2$ 3rd generation), 1.24 (d, *J* = 2.7 Hz, 9H, C-CH_3 1st generation, 2nd generation), 1.13 (s, 12H, C-CH_3 3rd generation). ^{13}C-NMR (100 MHz, $CDCl_3$): [*δ*, ppm] = 173.69, 171.94, 130.85, 128.61, 116.29, 98.27, 77.48, 77.16, 76.85, 66.10, 65.04, 46.97, 42.21, 25.57, 21.97, 18.62, 17.78. ESI-MS – $[M+Na]^+$ theo. 1248.59 Da, exp. 1248.56 Da. MALDI-MS - $[M+Na]^+$ theo. 1248.59 Da, exp. 1248.87 Da.

$(An)_4$-[G3]-Triazole-nonanyl acrylate (13b) was synthesized from $(An)_4$-[G3]-acetylene **7** and 9-azido-1-nonanyl acrylate **10b** according to the *general procedure for the Copper-Catalyzed Azide-Alkyne Cycloaddition (CuAAC)* to give **13b** as a colorless oil (70 %).

^{1}H-NMR (400 MHz, $CDCl_3$): [*δ*, ppm] = 7.71 (s, 1H, -N-*CH*=C), 6.39 (dd, *J* = 17.3, 1.5 Hz, 1H, CH_2=CH-), 6.11 (dd, *J* = 17.3, 10.4 Hz, 1H, CH_2=*CH*-), 5.81 (dd, *J* = 10.4, 1.5 Hz, 1H, CH_2=CH-), 5.26 (s, 2H, -CH_2-OCO-), 4.35 (t, *J* = 7.3 Hz, 2H, -CH_2-N), 4.31 - 4.18 (m, 12H, -C-CH_2-O- 1st generation, 2nd generation), 4.17 - 4.11 (m, 10H, O-CH_2-, -C-CH_2-O- 3rd generation), 3.62 (d, *J* = 12.2 Hz, 8H, -C-CH_2-O- 3rd generation), 1.95 - 1.84 (m, 2H, -CH_2-CH_2-N), 1.65 (dd, *J* = 14.3, 7.0 Hz, 2H, -O-CH_2-CH_2-), 1.41 (s, 12H, $C(CH_3)_2$ 3rd generation), 1.33 (d, *J* = 7.2 Hz, 22H, -$(CH_2)_5$-, $C(CH_3)_2$ 3rd generation), 1.24 (d, *J* = 3.1 Hz, 9H, C-CH_3 1st generation, 2nd generation), 1.13 (s, 12H, C-CH_3 3rd generation). ^{13}C-NMR (100 MHz, $CDCl_3$): [*δ*, ppm] = 173.66, 171.93, 142.01, 130.65, 128.74, 124.31, 98.24, 66.10, 65.02, 64.75, 58.73, 50.52, 46.86, 42.20, 39.43, 30.43, 29.26, 26.62, 26.00, 25.49, 22.04, 18.63, 17.78. ESI-MS – $[M+Na]^+$ theo. 1290.64 Da exp. 1290.64 Da. MALDI-MS - $[M+Na]^+$ theo. 1290.64 Da, exp. 1290.94 Da.

Polymerization Procedure

The macromonomer synthesis based on the dendronized acrylates **11-13** (see Scheme 7-5) proceeded with 10^{-3} mol·L^{-1} AIBN in a 5 wt% solution of hexyl acetate (3.8 g, 26.4 mmol) (freed from oxygen by purging with argon for about 40 minutes prior to the reaction) in a pressure tube at 140 °C. Specifically, the monomer (0.2 g, 5 wt%) was dissolved in hexyl acetate and degassed by purging with argon for about 20 min in a separate vial sealed airtight with a septum at ambient temperature. The initiator AIBN (10^{-3} mol·L^{-1}, its function is explained in more detail in the Results and Discussion section) was dissolved in hexyl acetate and added to the reaction mixture at 140 °C. After 5 min, the degassed monomer was added to the solvent and the pressure tube sealed airtight. The reaction solution was subsequently stirred for close to 72 h. After the reaction the solvent was removed under reduced pressure. The purity of the synthesized macromonomer was determined by (size exclusion chromatography) Electrospray Ionization Mass Spectrometry (SEC/ESI-MS).

Copolymerization Procedure

The synthesis of p(**11**)-*co*-EA MM employing the dendronized acrylate **11** proceeded in a 5 wt% solution of hexyl acetate (3.8 g, 26.4 mmol) (freed from oxygen by purging with argon for about 40 min prior to the reaction) in a pressure tube at 140 °C. Specifically, the monomer **11** (0.1 g, 2.5 wt%) was degassed by purging with argon for close to 20 min in a separate vial sealed airtight with a septum at ambient temperature. The same was applied to ethyl acrylate (0.1 g, 2.5 wt%). The initiator AIBN (10^{-3} mol·L^{-1}, its use is explained in more detail in the Results and Discussion section) was dissolved in hexyl acetate and added to the reaction mixture at 140°C. After 5 min the degassed monomer **11** and ethyl acrylate were successively added to the solvent and the pressure tube sealed airtight. The reaction solution was subsequently stirred for close to 72 h. Product isolation was performed analoguous to the (homo)macromonomer synthesis.

Hydrogenation Procedure

For the hydrogenation p(**11b**) MM was dissolved in a mixture of ethyl acetate and methanol 1/1 (v/v) and 20 mol% Pd/C was added to the solution. The reaction mixture was stirred close to 6 h at ambient temperature in an autoclave with a hydrogen pressure of 10 bar. The reaction mixture was subsequently flushed through a 20 µm PTFE membrane filter (VWR) to remove the residual catalyst and the solvent was removed under reduced pressure. The product purity was determined by Electrospray Ionization Mass Spectrometry (ESI-MS).

7.2.3 Results and Discussion

7.2.3.1 Dendronized Acrylates Synthesis

The synthesis and subsequent high-temperature polymerization of dendronized acrylates to yield dendronized vinylic macromonomers can be a versatile route towards brush-shaped polymers. Since acrylates undergo β-scission during polymerization resulting in MCRs,[9] it is possible to generate macromonomers, which can be used in further transformations due to their vinyl terminus. For the synthesis of the dendronized monomers acetylene-functionalized dendrons based on bis-MPA were synthesized according to the concepts of Hawker and co-workers.[252] Dendrons, from 1st to 3rd generation, were accomplished via a divergent growth approach, which combines esterification steps using acetonide protected bis-MPA anhydride and deprotection (activation) with DOWEX® resin. The acetylene cores were subsequently coupled to antecedently synthesized azido-functionalized acrylates via the copper-catalyzed azide-alkyne cycloaddition (CuAAC). The ligation reaction was catalyzed by copper(II)sulfate in the presence of sodium ascorbate yielding pure dendronized acrylates with a 1,4-disubstituted 1,2,3-triazole linkage after column chromatographic purification.

Scheme **7-6** depicts a detailed overview of the synthetic strategy including the synthesis of acetylene-terminated dendrons, azido-functionalized acrylates and the copper-catalyzed azide-alkyne cycloaddition (CuAAC) *click* reaction.

To enable an efficient as possible polymerization/oligomerization, i.e., macromonomer formation process,[11] a flexible spacer was introduced into the monomer between the dendron and the acrylic moiety to increase the spatial availability of the acrylic group. Two carbon spacers were introduced containing 6 and 9 CH_2-units, respectively. For the azide terminated acrylates 6-bromo-hexanol **8a** and 9-bromo-nonanol **8b**, respectively, was reacted with NaN_3 including a bromine-azide replacement followed by an esterification step with acryloyl chloride. After column chromatography with heptane:ethyl acetate, the acrylate precursors were isolated with yields of 81 % and 80 %, respectively. In the following modular ligation step, the dendronized acrylates were isolated in good yields up to 89 % after column chromatography. The dendronized acrylates were characterized via NMR spectroscopy and ESI as well as MALDI-ToF mass spectrometry. Figure 7-4 depicts a typical ^{1}H-NMR (a) and ESI-MS (b) spectrum of $(An)_4$-[G3]-triazole-hexyl acrylate (C_6G3).

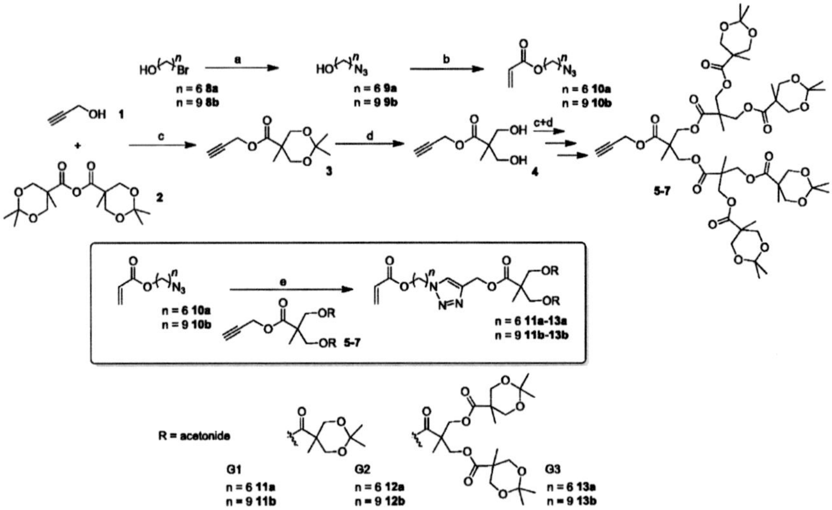

Scheme 7-6. Synthesis of the dendronized acrylates generation 1 to 3. Reaction conditions: (a) NaN_3, DMSO, reflux. (b) Acryloyl chloride, NEt_3, DMAP, CH_2Cl_2, rt. (c) DMAP, pyridine, CH_2Cl_2, rt. (d) DOWEX® H^+, methanol, 40 °C. (e) $CuSO_4 \cdot 5H_2O$, sodium ascorbate, THF:H_2O (3:1), rt.

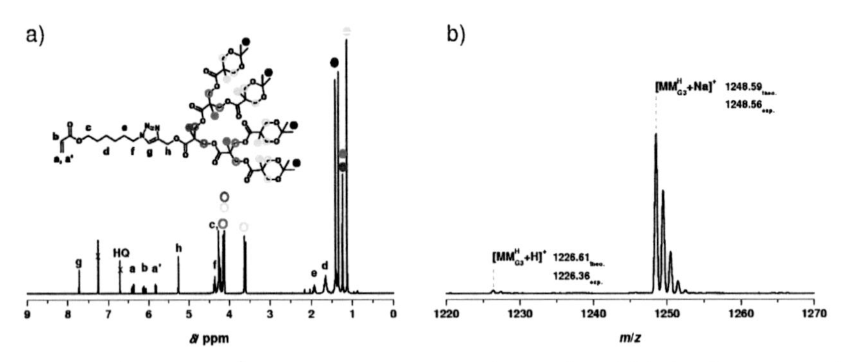

Figure 7-4. (a) 400 MHz ^{1}H-NMR spectrum measured in $CDCl_3$ and (b) ESI-MS spectrum of $(An)_4$-[G3]-triazole-hexyl acrylate (C_6G3) **13a**. Hydroquinone (HQ) was added after purification to prevent premature polymerization.

The success of the modular ligation step was confirmed by the appearance of the signal at 6.11 ppm related to the triazole proton (h). Further, a more detailed look at the chemical structure was achieved via ESI-MS measurements, with an assignment accuracy of $\Delta\, m/z = 0.03$ for $[MM^H_{G3}+Na]^+$. A similar result was achieved for $(An)_4$-[G3]-triazole-nonanyl acrylate (C_9G3) **13b** with a spacer length 9. ^{1}H-NMR (a) as well as ESI-MS (b) spectrum can be found in Figure 7-5.

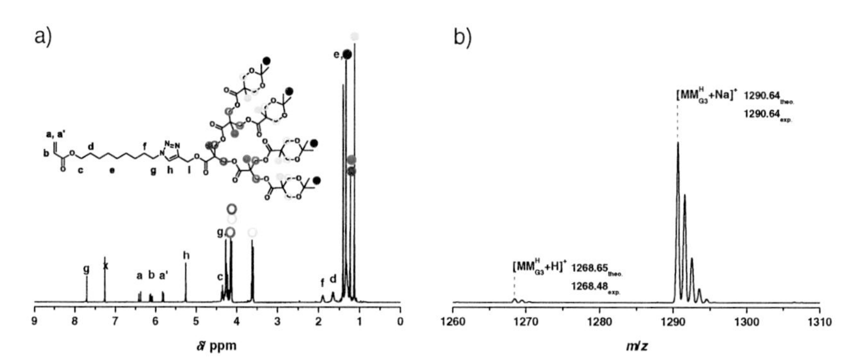

Figure 7-5. (a) 400 MHz ^{1}H-NMR spectrum measured in $CDCl_3$ and (b) ESI-MS spectrum of $(An)_4$-[G3]-triazole-nonanyl acrylate (C_9G3) **13b**.

The triazole signal (h) of the C_9G3 acrylate appears – compared to the C_6G3 acrylate – at slightly lower chemical shift at 5.81 ppm. The additional ESI-MS analysis evidences the ligation. The isotopic pattern of the C_9G3 acrylate is shifted by 42 Da to higher masses due to 3 additional CH_2-units of the implemented spacer. Apart from the sodium doped species, the charging with H^+ ions of the molecules $[M+H]^+$ is detectable in both spectra. To prevent the premature polymerization of the generated monomers hydroquinone was added to all reactions, as well as during storage of these compounds.

7.2.3.2 Macromonomer Formation via High-Temperature Acrylate Polymerization

In a significant extension of the previously synthesized macromonomer library,[11] the prepared dendronized acrylates were successfully polymerized via high-temperature polymerization to afford vinyl terminated oligomeric building blocks. The dendronized acrylates from generation 1 to generation 3, with a spacer length of 6 and 9 each (**11-13**), were deinhibited over neutral alumina prior to use. Auto-initiated high-temperature acrylate polymerization was carried out in hexyl acetate and small amounts of AIBN were added prior to polymerization.[9,11] It is hypothesized that AIBN does not initiate any polymerization activity as all initiator is decomposed within 0.15 minutes at 140 °C.[11,202] However, it removes minimal impurities in the system to generate products with high purity. Scheme 7-5 represents the general idea of the comb-like macromonomer formation based on dendronized acrylates from the 1st to 3rd generation.

In the current study, the high-temperature polymerization of the 1st generation, the 2nd generation and the 3rd generation dendron with a carbon spacer 6 and 9 each (**11-13**) was carried out. Below, the analytic process and results will be discussed on the example of the 1st generation with the carbon 9 spacer (**11b**). The corresponding analysis for higher generations and spacer lengths can be found in the Appendix. The successful polymerization process was initially evidenced by size exclusion chromatography as a convenient technique. Figure 7-6 presents the SEC chromatogram obtained from p(**11b**) MM. SEC chromatograms of p(**11-13**) MM can be found in the Appendix (Figure 13-1, p(**11a**) MM; Figure 13-3, p(**11b**)-*co*-EA MM; Figure 13-5, p(**12a**) MM; Figure 13-7, p(**13b**) MM).

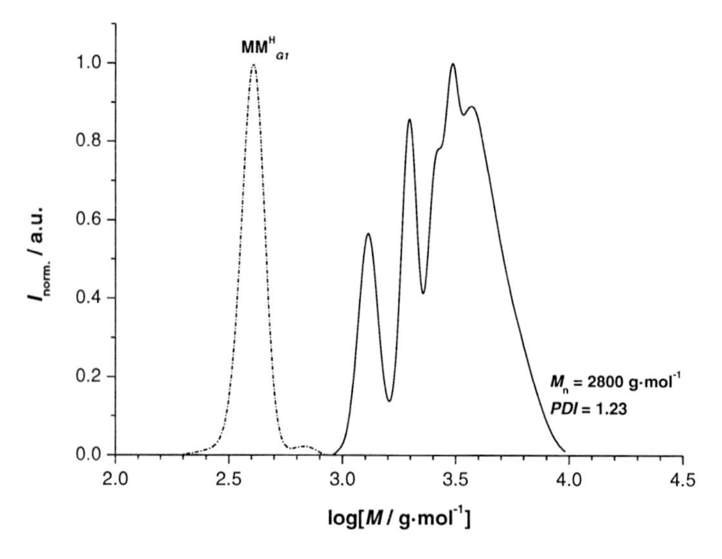

Figure 7-6. SEC chromatogram of p(**11b**) MM. The trace on the left hand side represents the residual monomer, whereas the trace on the right hand side shows the distribution after the polymerization.

Figure 7-6 displays two distributions: on the left hand side (dashed signal) the residual dendronized monomer is visible; on the right hand side the distribution of the macromonomer species is depicted. Both signals have been analyzed and normalized separately against polystyrene standards. Due to the relatively high molecular weight of the monomer (M(**11b**) = 451.27 g·mol^{-1}), the product distribution on the right hand side reflects the stepwise addition of the monomer to the growing polymer chain with well-resolved peaks. The system never reaches full conversion even at extended reaction times. Therefore a standard reaction time of 72 h for each polymerization was selected. Nevertheless, it is notable that the dendronized acrylates form macromonomers despite their bulky side chains, even for the 3rd generation species. With increasing dendron size, the formation of the macromonomers is limited to smaller numbers of repeat units in the final polymer. However, even in the 3rd generation material, the vinyl terminus is still detectable.

For p(**11b**) MM the number-average molecular weight is close to $M_n = 2800\ g{\cdot}mol^{-1}$, which translates to a degree of polymerization of $DP = 6.3$. The corresponding p(**13b**) MM (M(**13b**) = $1267.65\ g{\cdot}mol^{-1}$) shows a degree of polymerization of $DP = 3.4$ at a molecular weight of $M_n = 4300\ g{\cdot}mol^{-1}$ calculated from SEC. The tendency of the dendronized acrylates to form macromonomers with a large polymer backbone decreases with increasing dendritic generation of the monomer – as expected – due to the increasing steric hindrance during the polymerization process.

While the SEC analysis provides an indication that the dendronized acrylates polymerize indeed, a more detailed molecular proof is required. Thus, ESI-MS spectra were measured for all samples to analyze the macromonomers with respect to their endgroup fidelity and purity.[68,72] The molecular weight of the generation 1 and 2 macromonomers are well in the range of the calibrated m/z range of the ESI-MS experimental set-up while the 3rd macromonomers were analyzed via a hyphenated SEC/ESI-MS due to the high molecular weight of the monomer (M(**13a**) = $1225.60\ g{\cdot}mol^{-1}$, M(**13b**) = $1267.65\ g{\cdot}mol^{-1}$). Such a slice by slice ESI-MS analysis allows for a better ionization of higher molecular weight polymers in higher charge states, thus allowing an improved imaging of the material.

The synthetic procedure yields macromonomer species of the general structure as detailed in our previous study on acrylates-dented macromonomers.[11] During the polymerization one major product is formed, however, in general saturated and unsaturated species can be distinguished. The unsaturated products – the main synthetic target – with their vinyl terminus are capable to undergo further transformations to construct more complex macromolecules. The unsaturated species $\mathbf{MM^{H}}$ represents the main product of the synthesis; a separate species initiated via a transfer to solvent reaction, thus carrying a hexyl acetate radical fragment ($\mathbf{MM^{Hex}}$), is also identified. Both species carry a vinyl terminus and are thus equivalent in terms of their chemical reactivity. The saturated species $\mathbf{{}_{sat}P}$ and $\mathbf{{}_{sat}P^{Hex}}$ are only formed in small quantities.

Figure 7-7 depicts the ESI-MS spectrum of p(**11b**) MM as an example. Recall that the ESI-MS spectrum provides a number distribution. The spectrum represents the overall region from 400-3500 m/z where all signals associated with the main product $\mathbf{MM^{H}_{G1}}$ are assigned. The species $\mathbf{MM^{H}_{G1}}$ is related to the residual monomer found in the sample and species $\mathbf{MM^{H}_{nG1}}$ refer to increasing addition of monomers with increasing n.

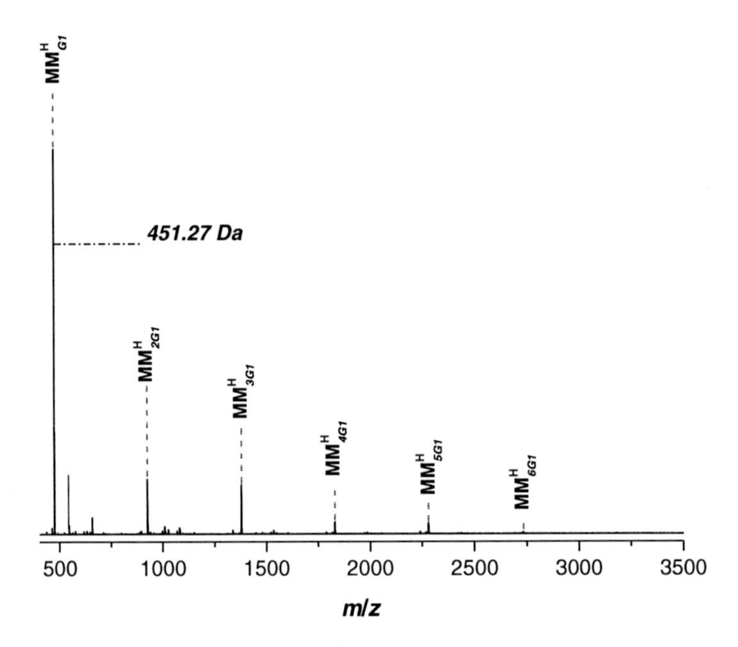

Figure 7-7. Overall ESI-MS spectrum of p(**11b**) MM. The polymer was synthesized via high-temperature acrylate polymerization in solution of hexyl acetate with 5 wt% monomer at 140 °C.

In Figure 7-8 a zoom spectrum of a repeat unit for a detailed assignment of all detected species is shown. The experimental and theoretical *m/z* values are summarized in Table 7-1 for all detected species. All mass assignments discussed in the following are accurate within ± 0.3 Da.

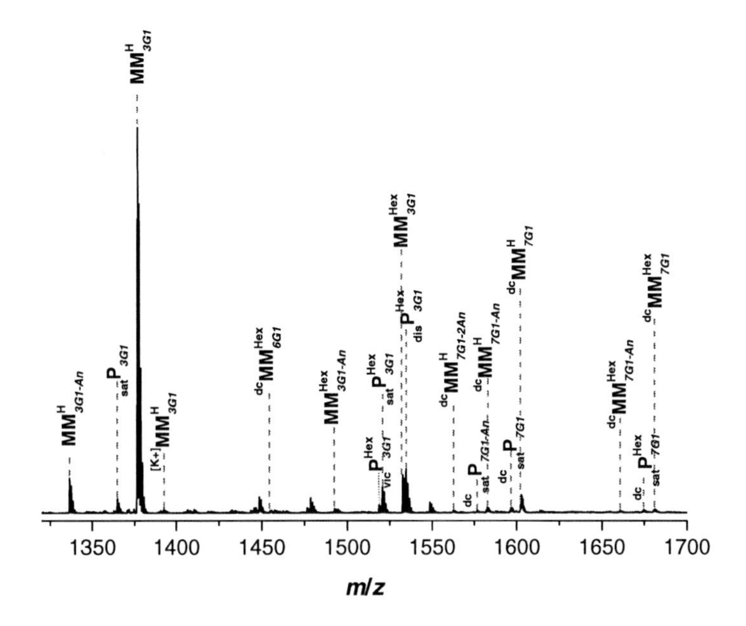

Figure 7-8. ESI-MS spectrum of p(**11b**) MM. A zoom spectrum of one repeat unit (451.27 Da) is depicted. The polymer was synthesized via high-temperature acrylate polymerization in solution of hexyl acetate with 5 wt% monomer at 140°C.

Inspection of Figure 7-8 shows as main product the $\mathbf{MM}^{H}_{nG1}$ species as well as $\mathbf{MM}^{Hex}_{nG1}$ in smaller quantities. The prefix to G1 represents the number of repeat units in the vinyl terminated polymer/oligomer. It should be noted that deprotection of the acetonide decorated surface of the dendron side chain takes place during the polymerization process (-40.03 Da for the acetonide protection in a single charged species). Fortunately, the deprotection ($\mathbf{MM}^{H}_{nG1\text{-}An}$) occurs only in small amounts and since the deprotected species are still capable to undergo further transformations due to their vinylic terminus, it is even conceivable to additionally use the deprotected hydroxyl functions for further transformation such as esterifications. Saturated products $_{sat}\mathbf{P}$ and $_{sat}\mathbf{P}^{Hex}$ are formed in very minor amounts. In the zoom spectrum, corresponding to a repeat unit of 3 in the polymer, double charged species are detectable, which evidence a propagation of the monomer up to 7 repeat units. As mentioned above, with increasing generation of the dendron segment in the monomer the propagation ability of the monomer decreases resulting in a lower degree of polymerization.

Table 7-1. Theoretical and experimental m/z ratios of the species of the p(**11b**) MM identified via the ESI-MS measurements. The resolution is close to 0.1 amu (see Figure 7-8). Listed below are the single charged sodium species $[M+Na]^+$ if not otherwise specified.

species	G1	- An	$m/z_{theo.}$	$m/z_{exp.}$	$\Delta\, m/z$
MM^H	3	1	1336.76	1336.72	0.04
$_{sat}P$	3	0	1364.79	1364.76	0.03
MM^H	3	0	1376.79	1376.72	0.07
MM^H	3	0	1392.77	1392.72	0.05
MM^{Hex} dc	6	0	1454.85	1454.72	0.13
MM^{Hex}	3	1	1492.88	1492.72	0.16
$_{vic}P^{Hex}$	3	0	1518.89	1518.76	0.13
$_{sat}P^{Hex}$	3	0	1520.91	1520.76	0.15
MM^{Hex}	3	0	1532.91	1532.76	0.15
$_{dis}P^{Hex}$	3	0	1534.93	1534.76	0.17
MM^H dc	6	2	1562.40	1562.32	0.08
$_{sat}P$ dc	7	1	1576.41	1576.20	0.21
MM^H dc	7	1	1582.41	1582.24	0.17
$_{sat}P$ dc	7	0	1596.43	1596.28	0.15
MM^H dc	7	0	1602.43	1602.28	0.15
$_{sat}P^{Hex}$ dc	7	1	1654.47	1654.28	0.19
MM^{Hex} dc	7	1	1660.47	1660.32	0.15
$_{sat}P^{Hex}$ dc	7	0	1674.49	1674.28	0.21
MM^{Hex} dc	7	0	1680.49	1680.32	0.17

To unambiguously evidence the vinyl terminus, hydrogenation of the sample was carried out. Figure 7-9 shows the zoom spectrum of p(**11b**) MM and its hydrogenated counterpart. The theoretical and experimental m/z ratios of the species are collated in Table 7-2.

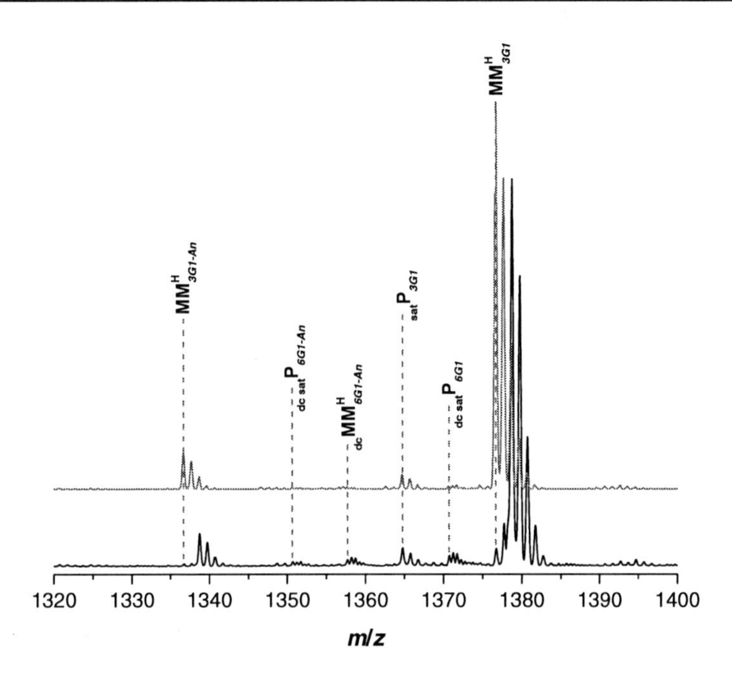

Figure 7-9. ESI-MS spectra of the vinyl terminated p(**11b**) MM (top) versus the hydrogenated polymer (bottom).

Table 7-2. Theoretical and experimental m/z ratios of the species of p(**11b**) MM and the hydrogenated analoga identified via the ESI-MS measurements. The resolution is close to 0.1 amu (see Figure 7-9). Listed below are the single charged sodium species $[M+Na]^+$ if not otherwise specified.

species	G1	- An	$m/z_{\text{theo.}}$	$m/z_{\text{exp.}}$	$m/z_{\text{theo.}}$ hydrogenated	$m/z_{\text{exp.}}$ hydrogenated	$\Delta\, m/z$
MM^H	3	1	1336.76	1336.72	1338.78	1338.72	0.06
$_{sat}P$ dc	6	1	1350.78	1350.72	1350.78	1350.68	0.10
MM^H dc	6	1	1356.78	1356.68	1357.79	1357.72	0.07
$_{sat}P$	3	0	1364.79	1364.76	1364.79	1364.76	0.03
$_{sat}P$ dc	6	0	1370.79	1370.72	1370.79	1370.72	0.07
MM^H	3	0	1376.79	1376.72	1378.81	1378.76	0.05

As can be clearly seen in Figure 7-9, the vinyl terminated species $\mathbf{MM^{H}_{3G1}}$, $\mathbf{MM^{H}_{3G1\text{-}An}}$ and $\mathbf{_{dc}MM^{H}_{6G1\text{-}An}}$ shift by 2 Da (1 Da for double charged species) to higher molecular weight, while all saturated species as $\mathbf{_{sat}P_{3G1}}$, $\mathbf{_{dc\ sat}P_{6G1}}$ and $\mathbf{_{dc\ sat}P_{3G1\text{-}An}}$ are not shifted. Thus, the chemical nature of the macromonomer species is fully established.

For dendron generation 2 and 3, similar product distributions can be found in the mass spectra. Further (SEC/)ESI-MS spectra can be found in the Appendix (Figure 13-2, p(**11a**) MM; Figure 13-4, p(**11b**)-*co*-EA MM; Figure 13-6, p(**12a**) MM; Figure 13-8, p(**13b**) MM).

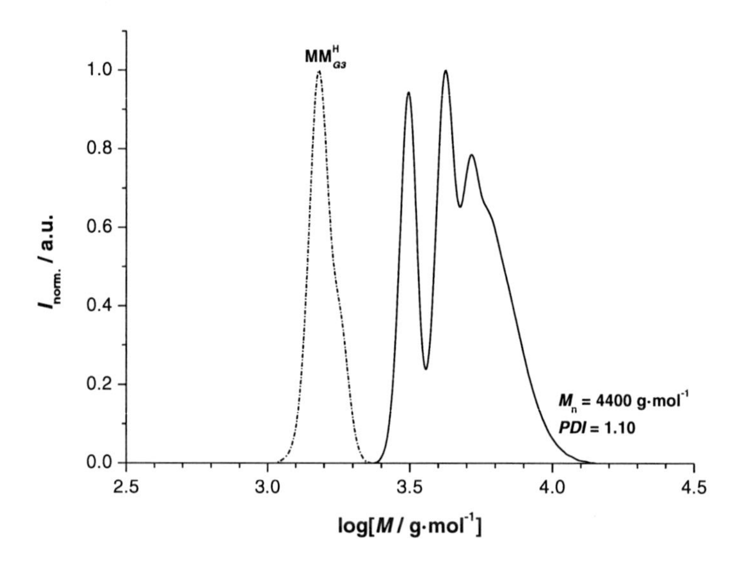

Figure 7-10. SEC chromatogram of p(**13a**) MM. The trace on the left hand side (dashed) represents the residual monomer whereas the trace on the right hand side shows the distribution after the polymerization. Both traces have been analyzed and normalized separately.

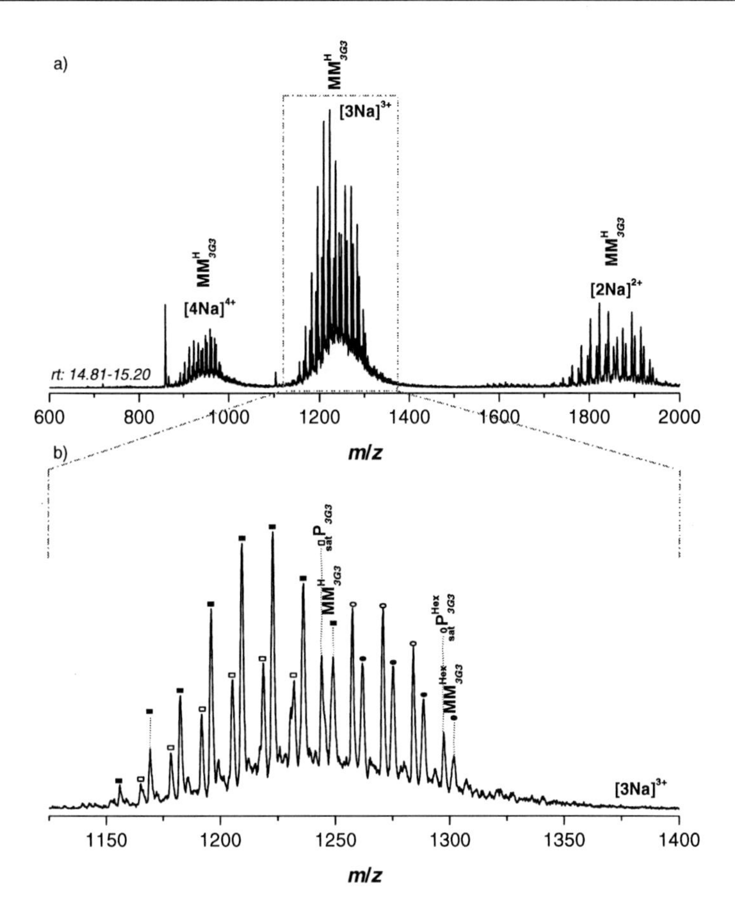

Figure 7-11. SEC/ESI-MS spectrum of p(**13a**) MM. The spectrum (a) represents a detailed spectrum at a specific retention time (14.81-15.20 min), whereas the spectrum (b) represents the zoom to the detected triple charged species at the above mentioned retention time. The polymer was synthesized via high-temperature acrylate polymerization in solution of hexyl acetate with 5 wt% monomer at 140 °C in an oxygen free atmosphere with 5×10^{-3} mol·L^{-1} AIBN.

Table 7-3. Theoretical and experimental *m/z* ratios of the species of p(**13a**) MM identified via the SEC/ESI-MS measurements. The resolution is close to 0.1 amu (see Figure 7-11). Listed below are the triple charged sodium species $[M+3Na]^{3+}$ if not otherwise specified.

species	G3	- An	$m/z_{theo.}$	$m/z_{exp.}$	$\Delta m/z$
MM^{H}	3	7	1155.87	1155.83	0.04
^{sat}P	3	6	1165.22	1165.00	0.22
MM^{H}	3	6	1169.22	1169.25	0.03
$_{sat}P$	3	5	1178.58	1178.33	0.25
MM^{H}	3	5	1182.58	1182.67	0.09
$_{sat}P$	3	4	1191.93	1191.92	0.01
MM^{H}	3	4	1195.93	1195.92	0.01
$_{sat}P$	3	3	1205.29	1205.25	0.04
MM^{H}	3	3	1209.29	1209.33	0.04
$_{sat}P$	3	2	1218.64	1218.58	0.06
MM^{H}	3	2	1222.64	1222.67	0.03
$_{sat}P^{Hex}$	3	5	1230.65	1230.58	0.07
$_{sat}P$	3	1	1231.99	1232.00	0.01
MM^{H}	3	1	1235.99	1235.92	0.07
$_{sat}P^{Hex}$	3	4	1244.00	1243.92	0.08
MM^{Hex}	3	4	1248.00	-	
MM^{H}	3	0	1249.35	1248.92	0.43
$_{sat}P^{Hex}$	3	3	1257.36	1257.33	0.03
MM^{Hex}	3	3	1261.36	1261.67	0.31
$_{sat}P^{Hex}$	3	2	1270.71	1270.67	0.04
MM^{Hex}	3	2	1274.71	1275.00	0.29
$_{sat}P^{Hex}$	3	1	1284.07	1284.00	0.07
MM^{Hex}	3	1	1288.07	1288.33	0.26
$_{sat}P^{Hex}$	3	0	1297.42	1297.33	0.09
MM^{Hex}	3	0	1301.42	1301.83	0.41

Table 7-4 shows an overview of the macromonomer content after each polymerization resulting from calculations of the measured ESI-MS spectra. Calculations were carried out analogous to the method used for the previously synthesized macromonomer library.[11]

Table 7-4. Percentage of macromonomer species resulting from the high-temperature acrylate polymerization, calculated from mass spectrometric data.

	MM [a]	MM^H [b]	An-protected [c]	MM^H [d]	M_n
		/ %			/ g·mol^{-1}
p(11a) MM	83.2	68.5	65.1	60.3	3000
p(11a)-*co*-EA MM	59.5	77.9	50.5	77.4	1900
p(11b) MM	81.7	91.4	82.1	83.5	2800
p(11b)-*co*-EA MM	69.9	77.7	66.5	81.1	1700
p(12a) MM	70.3	92.5	85.9	76.8	2900
p(13a) MM [e]	54.4	74.4	8.7	53.63	4400
p(13b) MM [e]	58.5	50.9	18.0	52.7	4300

[a,b] Amount of vinyl terminated polymer, containing b proton terminated macromonomers MM^H; [c,d] amount of acetonide-protected species, containing d proton terminated macromonomer MM^H; [e] values were calculated from SEC/ESI-MS spectra employing one repeating unit. Integration on other retention volumes provides similar results, however, the error associated with the data given for p(**13**) MM is significantly higher than for the other macromonomers.

Column 1 depicts the content of macromonomer species with vinyl terminus in the sample. Column 2 contains vinyl terminated polymers with a proton terminus $\mathbf{MM^H}$ related to the overall amount of macromonomeric species with a vinyl terminus. Vinyl terminated polymers as useful synthetic building blocks are built in acceptable yields exceeding 70 % except for 3rd generation macromonomeric products. The values for the p(**13**) MM were calculated from SEC/ESI-MS spectra (refer to the provided retention time of the sample shown in Figure 7-11 and Figure 13-8 in the Appendix).

In each sample, the $\mathbf{MM^H}$ species occurs as the main product and all mentioned side products are only detectable in small quantities. However, it has to be noted that the SEC/ESI-MS spectra for the 3rd generation are slightly deviant from the ESI-MS spectra of the lower generations. Due to the higher sensitivity of the ESI-MS after the pre-separation via SEC the side products are better visible for each retention time. However, the overall (low) integral amount of side products is identical between ESI-MS and SEC/ESI-MS. Compared to the 1st and 2nd generation macromonomers, deprotection of the acetonide surface occurs to a slightly higher extent, which affects the purity of protected species during the polymerization process. For the 3rd generation protected species up to 18 % are present, whereas the amount of acetonide protected species is 51 % to 86 %, depending on generation and spacer length. Even for the species $\mathbf{MM^{H}_{G3}}$, representing the residual monomer, the deprotection of the acetonide surface is higher compared to lower generations. It is assumed that the higher the generation of the dendrons, the higher deprotection occurs. At high temperature deprotection occurs to a certain extent, i.e., for a 3rd generation acrylate 8 hydroxy functionalities are protected with 4 acetonide (An) protection groups. Thus, 4 deprotections occur – at maximum – for each repeat unit. The quantification of the deprotected species is beset with some uncertainty since deprotection could occur on each repeat unit in the system. For example, in the $\mathbf{MM^{H}_{3G3\text{-}4An}}$ macromonomer with 3 repeat units 4 deprotections took place (m/z = 1195.93). These deprotections could have taken place partly on each repeat unit or – alternatively – completely deprotected one entire monomer unit. Note that the deprotected species still carry a vinylic terminus.

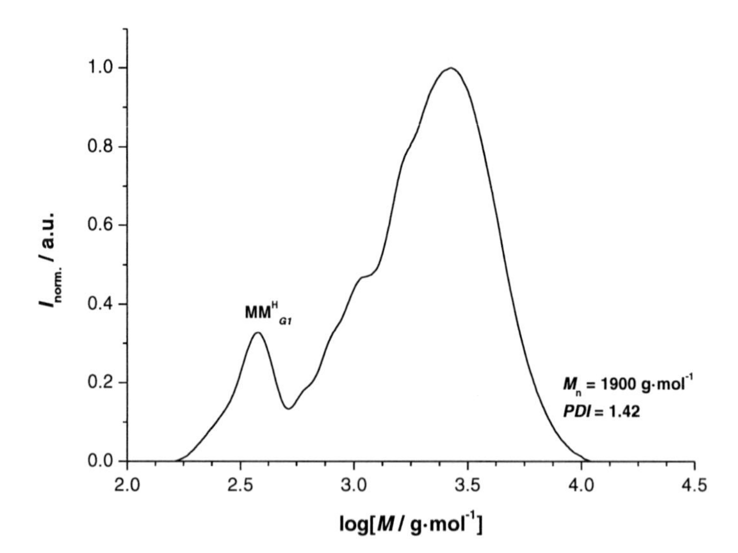

Figure 7-12. SEC chromatogram of p(**11a**)-*co*-EA MM. The peak on the left hand side represents the residual monomer, whereas the shifted part on the right hand side corresponds to the distribution of the copolymerization product.

While it is pleasing to note that the dendritic acrylates undergo transformations to macromonomer species, it is mandatory to establish whether copolymeric macromonomers can also be generated. For this purpose the copolymerization behavior of the acrylic dendrons **11** with ethyl acrylate was explored. In the following the copolymerization of **11a** with ethyl acrylate will be discussed in detail. The analytical results for the dendronized acrylate **11b** can be found in the Appendix. A ratio of 1:1 for the co-monomers dendritic acrylate (**11**) : ethyl acrylate has been applied in a 5 wt% solution in hexyl acetate as for the homopolymerization. The SEC trace in Figure 7-12 shows a clear shift to higher molecular weights, which proofs the presence of a polymerization process; residual monomer is also detectable.

To establish whether the generated polymeric material is truly a macromonomeric copolymer, a molecular assessment via ESI-MS is required. Figure 7-14 shows a typical ESI-MS spectrum of p(**11a**)-*co*-EA MM with the overall spectrum from 400–2000 *m/z* (above) and a specific zoom spectrum to a repeat unit of ethyl acrylate (below) (assignments are listed in Table 13-2 in the Appendix section).

The main distribution in the spectrum can clearly be assigned to $\mathbf{MM^{H}_{G1+n}}$, i.e., the copolymer with one dendronized monomer as repeat unit followed by *n* ethyl acrylate repeat units (100.05 Da). The second intense distribution corresponds to a deprotected species $\mathbf{MM^{H}_{G1\text{-}An+n}}$. The achievable amount of saturated species is higher compared to the high-temperature polymerization of pure macromonomer. Since deprotection occurs for the 1st generation acrylate, 3 different monomers acting as separate co-monomers are present in the system: protected dendronized acrylate, deprotected dendronized acrylate and ethyl acrylate (see Figure 7-13).

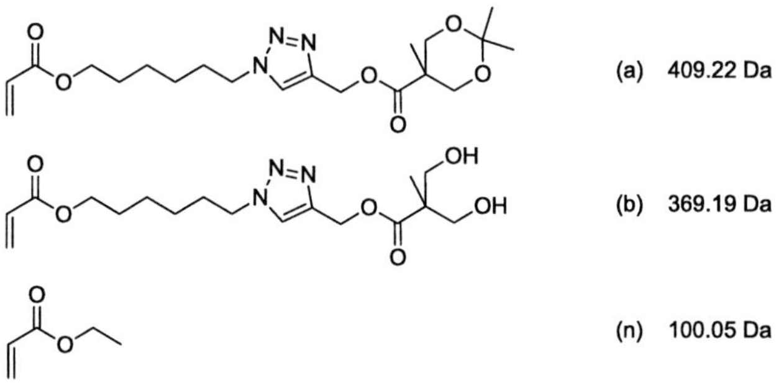

(a) 409.22 Da

(b) 369.19 Da

(n) 100.05 Da

Figure 7-13. Chemical structures of the monomers occurring in the copolymerization process of **11a** with ethyl acrylate.

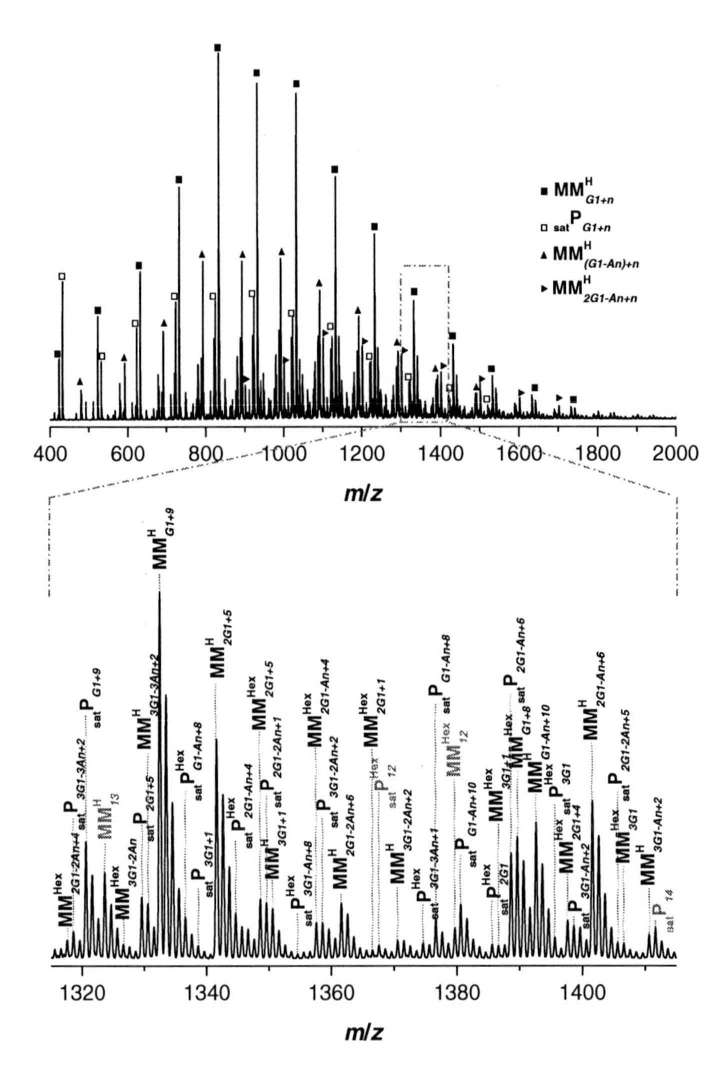

Figure 7-14. ESI-MS spectrum of p(**11a**)-*co*-EA MM. The overall spectrum (above) and the zoom spectrum of a repeat unit (100.05 Da EA) (below) are depicted. The polymer was synthesized via high-temperature acrylate polymerization in solution of hexyl acetate with 2.5 wt% **11a** and 2.5 wt% EA at 140 °C in an oxygen free atmosphere with 5×10^{-3} mol·L^{-1} AIBN.

The fact that three monomers are present in the mixture leads to a large variety of species found in the ESI-MS spectrum. The calculation of the macromonomer content from ESI-MS measurements indicate a decrease in macromonomeric species compared to the homopolymerization of the dendronized acrylates. Table 7-4 indicates a macromonomer content of 83 % for p(**11a**) MM and 60% for the copolymer p(**11a**)-*co*-EA MM, respectively. Even in the 2nd generation the formation of vinyl terminated polymers is slightly lower for the copolymer p(**11b**)-*co*-EA MM (70 %) compared to the homopolymer p(**11b**) MM with 82 % macromonomer. Since the copolymerization with a small acrylate as a backbone spacer does not result in a more efficient synthesis of well-defined macromonomers with a higher molecular weight, it can be concluded that the copolymerization itself is a versatile route to create molecules with different chemical and physical properties. However, it is recommended to use the dendronized acrylates in homopolymerization processes to obtain as pure as possible dendronized macromonomers.

The role of the size of the side chain spacer between the acrylic moiety and the dendrons cannot easily be established. Even though the influence of the amount of generations of the dendron on the propagation during the polymerization is clearly evidenced, the carbon spacer between the acrylic moiety and the dendron does not affect the polymerization process. Thus, the spacer length chosen for the monomer synthesis are too similar to impact a significant influence on the polymerization in the high-temperature acrylate synthesis of macromonomers.

7.2.4 Conclusions

Dendronized acrylates based on bis-MPA have successfully been synthesized via copper-catalyzed azide-alkyne cycloaddition (CuAAC). The benign *click* reaction gives rise to the combination of an acrylic functionality with well-defined dendritic wedges. The monomers were analyzed via NMR, ESI-MS and MALDI-ToF. These dendronized acrylates from 1st to 3rd generation, containing an alkyl chain spacer, were subjected to the auto-initiated high-temperature acrylate synthesis forming vinyl terminated polymers. For a better availability of the polymerizable group two different flexible spacers, carbon 6 and 9, have been introduced into the dendronized monomer structures between the acrylic moiety and the dendronized structures. The polymerization was performed at 140 °C in a 5 wt% solution of hexyl acetate with a 2,2'-azobis(isobutyronitrile) (AIBN) concentration of 5×10^{-3} $g \cdot mol^{-1}$. In addition, the homopolymerization of the molecules and the copolymerization with ethyl acrylate was carried out. The macromonomers were analyzed via SEC and (SEC/)ESI-MS. The achievable number-average molecular weight, M_n, was between 1700 and 4400 $g \cdot mol^{-1}$. The degree of polymerization, DP_n, decreases with increasing generations of the dendronized acrylates from 6.3 to 3.4. The approach of subjecting these monomers to the high-temperature acrylate synthesis thus constitutes a feasible avenue for generating versatile synthetic building blocks for further macromolecular transformations.

8

A Detailed Investigation of the Free Radical Copolymerization Behavior of *n*-Butyl Acrylate Macromonomers

8.1 Introduction to Free Radical Copolymerization

A conventional free radical polymerization of one specific monomer type gives rise to homopolymers. More often, however, industrial applications demand a combination of the properties of various homopolymers. The copolymerization of at least two monomer types virtually enables the synthesis of an unlimited variety of products. A simultaneous polymerization of two monomer types is commonly termed a binary copolymerization. More generally, the expression 'multi-component copolymerization' is used for a (radical) polymerization of a mixture of monomers.

Depending on the synthetic pathway and reactivity of the monomers, several distributions and architectures of copolymers are feasible. The most common ones are (a) statistical copolymers (incorporation of the monomers occurs randomly), (b) alternating copolymers, (c) block copolymers featuring long monomer sequences and (d) graft copolymers with branching from the polymer backbone. The above mentioned structures of copolymers are depicted in Figure 8-1.

Parts of this chapter were reproduced with permission from John Wiley & Sons.[13]

Already a binary copolymerization allows the formation of a large variety of products only by changing the nature, relative amount, and architecture of two monomers within the polymerization mixture. The fundamental aspects of a binary copolymerization (often simply referred to as 'copolymerization') are detailed in the following section.

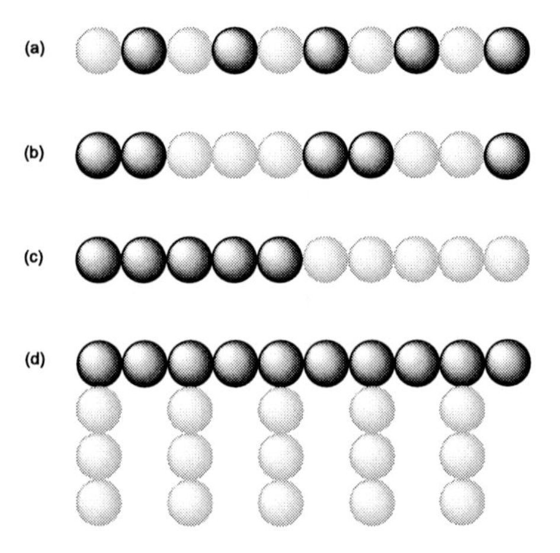

Figure 8-1. Schematic of possible copolymers of various compositions. (a) statistical, (b) alternating, (c) block and (d) graft copolymer.

8.2 Kinetics of Free Radical Copolymerization

The challenge of the synthesis of well defined materials and unlimited product varieties is associated with a thorough understanding and strict control of the copolymerization process.

8.2.1 The Terminal Model

The copolymerization kinetics of a binary copolymerization mixture may be described in a simple approach via the so-called *terminal model*, which only considers the terminal repeat unit of a propagating macro-radical to be relevant for the copolymerization kinetics.[263] The *terminal model* traces back to Mayo and Lewis as well as Alfrey and Goldfinger who first introduced the concept in the 1940s.[263-265] In most copolymerizations the *terminal model* adequately describes the composition data, while a joint description of both rate and composition data often fails. The basis of the *terminal model* is represented by the following equations (8-1) to (8-4)

$$\text{R-M}_1\cdot + \text{M}_1 \xrightarrow{k_{11}} \text{R-M}_1\text{M}_1\cdot \quad (8\text{-}1)$$

$$\text{R-M}_1\cdot + \text{M}_2 \xrightarrow{k_{12}} \text{R-M}_1\text{M}_2\cdot \quad (8\text{-}2)$$

$$\text{R-M}_2\cdot + \text{M}_1 \xrightarrow{k_{21}} \text{R-M}_2\text{M}_1\cdot \quad (8\text{-}3)$$

$$\text{R-M}_2\cdot + \text{M}_2 \xrightarrow{k_{22}} \text{R-M}_2\text{M}_2\cdot \quad (8\text{-}4)$$

with a rate coefficient k_{pij}, where i,j = 1 or 2. For a binary mixture with monomer M_1 and monomer M_2 as well as two propagating species, 4 propagation reactions are possible.[266-267] It is assumed that the propagation proceeds irreversibly, potential side reactions during the polymerization are not considered further. The propagating radical R-$M_1\cdot$ adds the monomer M_1 with a rate coefficient k_{11} (8-1) and the monomer M_2 with a rate coefficient k_{12} (8-2) and so on. The propagation steps (8-1) and (8-4) are referred to as homo-propagation (the monomer is adding to a propagating radical of the same monomer) whereas the propagation steps (8-2) and (8-3) are termed cross-propagation (the alternate monomer is added). The reactivity ratios of the participating co-monomers in the polymerization are defined as the ratios of the rate coefficient $\frac{k_{11}}{k_{12}} = r_1$ and $\frac{k_{22}}{k_{21}} = r_2$, respectively.[268] Knowledge of reactivity ratios can be employed to determine the (instantaneous) copolymer composition in a binary system. The utilization of simple mathematics including a steady-state assumption and mass balance during the copolymerization yields the copolymerization equation (8-5).

$$\frac{d[M_1]}{d[M_2]} = \frac{[M_1]\ (r_1[M_1]+[M_2])}{[M_2]\ ([M_1]+r_2[M_2])} \quad (8\text{-}5)$$

In equation (8-5) the copolymer composition ($d[M_1]/d[M_2]$) represents the molar ratio of the two monomers in the copolymer. The molar ratio is related to the monomer feed ratio $[M_1]/[M_2]$ and the reactivity ratios r_1 and r_2 of the respective monomers. The copolymerization equation (8-5) is also known as the Mayo-Lewis equation. The knowledge of the ratio of the rate coefficients – the reactivity ratios – provides the probability of two monomers to undergo copolymerization.

The determination of the reactivity ratios is based on experimental copolymer composition data which are evaluated graphically. Commonly, a copolymerization equation expressed by mole fractions is used for this purpose. The copolymerization equation (8-5) which is related to monomer concentrations needs to be further adjusted for the reactivity ratio determination. In principle, the determination of the reactivity ratio of the copolymerization can be accomplished by further calculations based on the Mayo-Lewis equation (8-5). The evaluation of the reactivity ratios follows from experimental copolymer composition data. Various co-monomer feeds f are applied to calculate reactivity ratios from the resulting low conversion copolymer composition. Once the mole fraction F of one component in the copolymer is determined, the values can be used in the Mayo-Lewis equation for copolymers.[265,269-270]

$$\frac{F_1}{F_2} = \frac{d[M_1]}{d[M_2]} = \frac{f_1 \cdot (r_1 \cdot f_1 + f_2)}{f_2 \cdot (r_2 \cdot f_2 + f_1)} \tag{8-6}$$

$$F_1 = \frac{r_1 \cdot f_1^2 + f_1 \cdot f_2}{r_1 \cdot f_1^2 + 2 \cdot f_1 \cdot f_2 + r_2 \cdot f_2^2} \tag{8-7}$$

with the mole fraction of component i in the monomer mixture f_i, the mole fraction of component i in the copolymer F_i and the reactivity ratio $r_i = k_{pii}/k_{pij}$ with $i = 1,2$. As equations (8-6) and (8-7) are differential equations, the *r*-values are used to calculate the copolymer composition in a system with two monomers at infinitely low conversions.

Consider, however, that the *terminal model* simplifies the actual copolymerization; several research groups have observed deviations from theoretical expectations.[271] Firstly, Fukuda and co-workers critically tested the *terminal model* on the example of the copolymerizations of styrene and methyl methacrylate, and methyl acrylate and *p*-chlorostyrene, respectively.[272-273] As a consequence the experimental observations differed from the theoretical calculations whereupon a model considering the influence of penultimate units was suggested for the description of the copolymerization processes.

8.2.2 Penultimate Model

The so-called *penultimate unit effect model* was shown to be superior to the *terminal model* in the work of Fukuda,[272] Davis[274] and Coote and co-workers.[275-276] The penultimate model – in comparison to the *terminal model* – considers the penultimate units on the chain of the growing macro-radicals.[277-278] This consideration implies 8 propagation steps described by 4 reactivity ratios.[266] The extended model describing copolymer processes gives more accurate reactivity ratios yet the *terminal model* remains an attractive 'engineering-type' option for copolymer analysis.

8.3 Determination of the Reactivity Ratios based on the Terminal Model

Based on the Mayo-Lewis equation (8-5) several options for the graphical determination of reactivity ratios are known in the literature. The most often used are discussed below.

8.3.1 Evaluation based on Mayo and Lewis

For the graphical evaluation of experimental copolymerization composition data various methods have been introduced. A linear form of the copolymer equation (8-5) was introduced by Mayo and Lewis in 1944,[265] i.e.,

$$r_2 = \frac{[M_1]}{[M_2]}\left[\frac{d[M_1]}{d[M_2]}\left(1+\frac{r_1[M_1]}{[M_2]}\right)-1\right] \tag{8-8}$$

The applied mole fractions f and the copolymer compositions F are inserted into equation (8-8). For the remaining unknown quantity r_1 various values are assumed and subsequently plotted against r_2. The intersection of various straight lines gained from the experimental data sets features the best value for r_1 and r_2. Additionally, this evaluation can also carried out via a linear-squares regression analysis.[19]

8.3.2 Further Evaluation Methods

In the following years after the work of Mayo and Lewis the determination of reactivity ratios in copolymerization processes was further developed by several other researchers. The basis for linear evaluation of experimental data was still based on the *terminal model* of copolymerization and thus refined for the determination of reactivity ratios. In 1950 Fineman and Ross rearranged equation (8-8) to generate a straight line with a slope r_1 and intercept r_2.[279] Kelen and Tüdos further developed the linear evaluation in 1975.[280] For a detailed discussion of the above mentioned methods the reader is referred to literature.[281-283]

8.4 Types of Copolymerization Behavior

According to the values of the monomer reactivity ratios different copolymerization behaviors are observed. All observed copolymerization behaviors are categorized by their mathematical product $r_1 \times r_2$ which ranges from zero to unity. Hereinafter, a short overview will be given.

8.4.1 Ideal Copolymerization $r_1 \times r_2 = 1$

The term ideal copolymerization is employed when the product of both reactivity ratios equals unity. In this case, the addition of a monomer, either M_1 or M_2, to the propagating radical shows no preference and follows a random (Bernoullian) process. The terminal monomer of the propagating radical does thus not influence the propagation process.

Statistical copolymers are obtained in this case. In a special case when both reactivity ratios, r_1 and r_2, equal unity the copolymer composition F_1 equals the actual monomer feed f_1. This ideal copolymerization behavior is represented by the angle bisector (dotted line) in Figure 8-2.

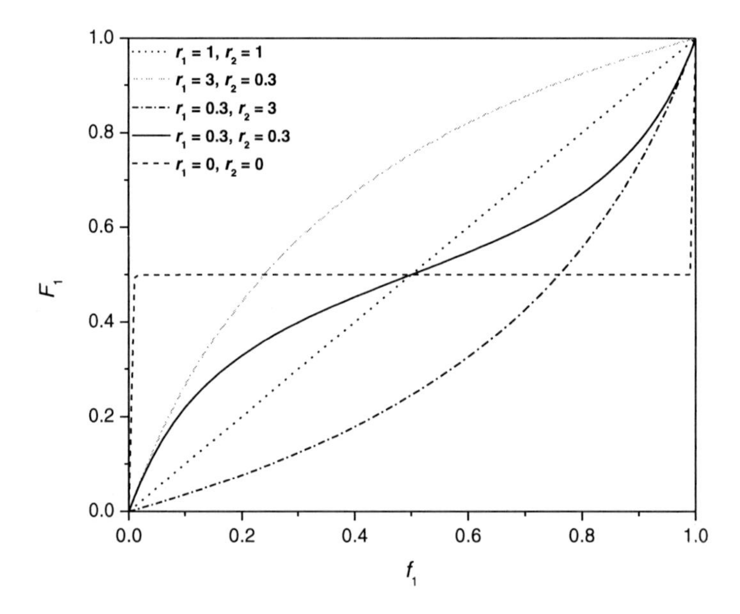

Figure 8-2. Types of copolymerization behavior. The copolymer composition F_1 of the monomer M_1 is plotted against the mole fraction in the monomer feed f_1. Ideal copolymerizations (dotted, gray and black dotted-dashed line) and alternating copolymerizations (straight and dashed line) are shown.

If the product of the reactivity ratios r_1 and r_2 is unity, the *r*-values can either be smaller or greater than unity, i.e., $r_1 < 1$ and $r_2 > 1$ or vice versa. In Figure 8-2 this behavior is depicted by the gray dotted-dashed line for $r_1 = 3$ and $r_2 = 0.3$ and the black dotted-dashed line for $r_1 = 0.3$ and $r_2 = 3$ as representative examples. The copolymerization process still occurs randomly, however, with longer blocks of the more reactive monomer towards the propagating radical.

8.4.2 Alternating Copolymerization $r_1 \times r_2 = 0$

On the other extreme, the product of both reactivity ratios, r_1 and r_2, equals zero. In this case the copolymerization equation is reduced to $F_1 = 0.5$ which implies that the amount of

incorporated monomer to the copolymer is independent of the employed co-monomer feed f_1 (see dashed line Figure 8-2). The propagating radical preferentially adds the other monomer to the radical terminus (cross-propagation) and consequently alternating copolymers are formed.
The copolymerization behavior of most polymers is found to be between ideal and alternating copolymerization.

8.4.3 Copolymerizations $r_1 \neq r_2$

Copolymerizations with unequal reactivity ratios r_1 and r_2 can be separated into two copolymerization regimes, giving different copolymeric structures.

For reactivity ratios both smaller than unity, i.e., $r_1 < 1$ and $r_2 < 1$, the copolymerization plot (see straight line Figure 8-2) crosses the line of the ideal copolymerization (dotted line) where $F_1 = f_1$. This type of copolymerization is referred to as alternating azeotropic copolymerization. The cross-propagation is favored forming alternating copolymers. On the crossover point or azeotropic point no change in feed concentration occurs which means that the monomer incorporated into the copolymer is the same as in the feed co-monomer mixture. The azeotropic point can be located at any position of the angle bisector.

Conversely, when both reactivity ratios are greater than unity ($r_1 > 1$ and $r_2 > 1$) block copolymer-like structures are formed. The probability of the propagating radical adding to the same monomer is higher than the probability of adding to the alternate monomer; thus homo-propagation is favored over cross-propagation. In the resulting copolymer longer chains of each monomer are found, tending more towards a block copolymer. This behavior does occur, albeit rarely, in some conventional free radical copolymerizations.

8.5 The Free Radical Copolymerization of *n*-Butyl Acrylate Macromonomer

The vinyl terminus of the macromonomers can be employed in copolymerizations with a conventional monomer; this strategy represents an efficient pathway for constructing branched polymer architectures. Several research groups have employed free radical polymerization as well as living/controlled radical polymerization methods such as ATRP and NMP for the copolymerization of post-functionalized macromonomers.[29,124,270,284-289]

In the current chapter the free radical copolymerization of macromonomers synthesized via the one-pot – one-step synthesis during high temperatures ($T \geq 140$°C) is addressed. It will be focused on *n*-butyl acrylate macromonomers (BAMM) copolymerized with benzyl acrylate (BzA) as a co-monomer. Scheme 8-1 gives an overview of the synthetic route employed.

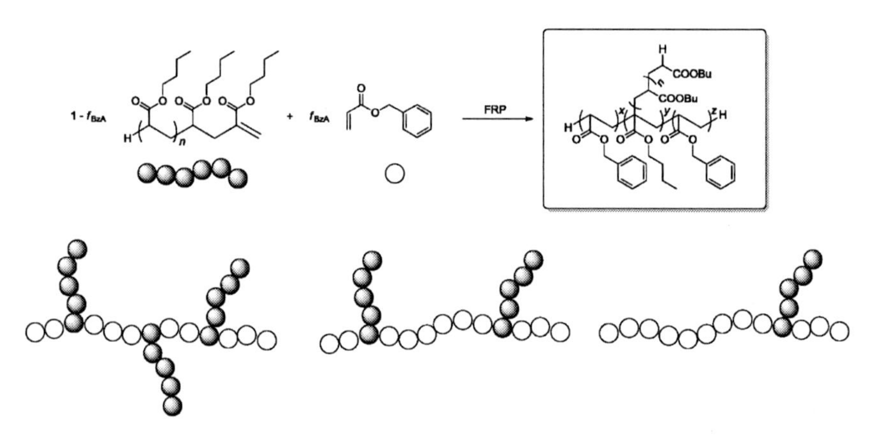

Scheme 8-1. Synthesis of copolymers via the macromonomer approach. BAMM was polymerized with benzyl acrylate (BzA) as a co-monomer.

As an initial analysis method size exclusion chromatography (SEC) was employed for the molecular weight determination of the generated copolymers. For the determination of the reactivity ratios via the copolymer composition ^{1}H-NMR spectroscopy and integration of specific signals were utilized. However, conventional SEC analysis of copolymers gives an indication regarding the formation of high molecular weight species, but is not specified with regard to the composition of the copolymer in terms of block length of the co-monomers and the ratio of secondarily build homopolymer, since homo- and copolymers of the same hydrodynamic volume cannot be distinguished. The use of liquid chromatography at critical conditions (LCCC), as an additional characterization method, can yield more detailed information.

The analysis of the statistical copolymers pBAMM-*co*-pBzA synthesized from BAMM with BzA via free radical copolymerization should evidence the copolymer structure of the samples as well as the quantity of potentially remaining – even after purification – homopolymer.

8.5.1 Synthesis

*Synthesis of n-Butyl Acrylate Macromonomer **BAMM***

For the macromonomer synthesis *n*-butyl acrylate was polymerized with 5×10^{-3} mol·L^{-1} AIBN in a 5 wt% solution of hexyl acetate (285 g, 1.98 mol) (freed from oxygen by purging with argon for about 40 minutes prior to the reaction) in a 2-neck flask at 140 °C. The monomer (15.0 g, 0.12 mol, 5 wt%) was degassed by purging with argon for about 20 minutes in a separate vial sealed airtight with a septum at ambient temperature. The initiator AIBN (5×10^{-3} mol·L^{-1}) was dissolved in hexyl acetate, freed from oxygen by purging with argon and added to the reaction mixture at 140 °C. After 5 minutes the degassed monomer was added to the solvent. The reaction solution was subsequently stirred for close to 10 hours. After the reaction, the solvent was removed in a vacuum oven at 45 °C. The purity of the synthesized macromonomer was determined by Size Exclusion Chromatography (SEC) and Electrospray Ionization Mass Spectrometry (ESI-MS). **BAMM** – M_n = 2300 g·mol^{-1}, *PDI* = 1.90; the purity of BAMM exceeds 92 % as estimated by ESI-MS.

*Copolymerization of n-Butyl Acrylate Macromonomer **pBAMM-co-pBzA***

For the synthesis of the copolymers pBAMM-*co*-pBzA variable mole fractions (0.1 to 0.9) of the monomer benzyl acrylate were used. The reaction preparation was performed in an inert gas atmosphere inside a glove box. Benzyl acrylate (f_{BzA}: 0.15 mol·L^{-1} for f_{BzA} = 0.1 up to 1.21 mol·L^{-1} for f_{BzA} = 0.9) was added to a solution of BAMM (f_{BAMM} = 1-f_{BzA}: 1.33 mol·L^{-1} for f_{BAMM} = 0.9 down to 0.13 mol·L^{-1} for f_{BAMM} = 0.1) and AIBN (1 × 10^{-1} mol·L^{-1}) in 1 mL toluene with an Eppendorf pipette. The reaction mixture was then sealed airtight with a septum and stirred at 60 °C to a conversion of approximately 40 %. The polymerization was stopped by cooling the mixture with liquid nitrogen. The solution was dried in a vacuum oven at 45 °C to remove residual monomer. The resulting copolymer was recovered by precipitating the crude copolymer three times in an excess of cold methanol and then dried *in vacuo*. For a detailed analysis of the obtained statistical graft copolymers refer to the Results and Discussion section.

*Synthesis of n-Butyl Acrylate Standards **pBA***

Polymer standards for LCCC measurements were synthesized via the RAFT technique. The initial ratio of $[BA]_0$: $[DoPAT]_0$: $[AIBN]_0$ was 310 : 1 : 0.2. The reaction was performed at 60 °C for variable time intervals. The polymerization was stopped by cooling the mixture with liquid nitrogen. The residual monomer was subsequently removed in a vacuum oven at 45 °C. The polymers were analyzed via SEC and SEC/ESI-MS. The specifics for the four pBA standards are as follows: **pBA1** – M_n = 8100 g·mol^{-1}, *PDI* = 1.09; **pBA2** – M_n = 12100 g·mol^{-1}, *PDI* = 1.06; **pBA3** – M_n = 19000 g·mol^{-1}, *PDI* = 1.04; **pBA4** – M_n = 24600 g·mol^{-1}, *PDI* = 1.04.

8.5.2 Results and Discussion

8.5.2.1 Copolymer Synthesis

The determination of the copolymer structure of copolymers prepared during free radical polymerization of macromonomers synthesized via the high-temperature acrylate polymerization with an acrylic monomer requires the synthesis of variable copolymer samples. Thus, the copolymerization of BzA and BAMM was carried out for several initial co-monomer feeds in solution of toluene with AIBN as a source of radicals. Polymerization in bulk is impractical due to diffusion restrictions associated with a decrease in the f_{BzA} fraction since the viscosity of the pure BAMM is relatively high. The addition of a solvent to the polymerization mixture of BzA and BAMM decreases viscosity. The mole fraction f_{BzA} was varied from 0.1 to 0.9 in 0.1 increments. After approximately 40 % conversion, determined gravimetrically, the copolymerization was stopped and residual macromonomer was removed from the crude copolymer by threefold precipitation in an excess of cold methanol. A typical SEC elugram of pBAMM-*co*-pBzA is shown in Figure 8-3.

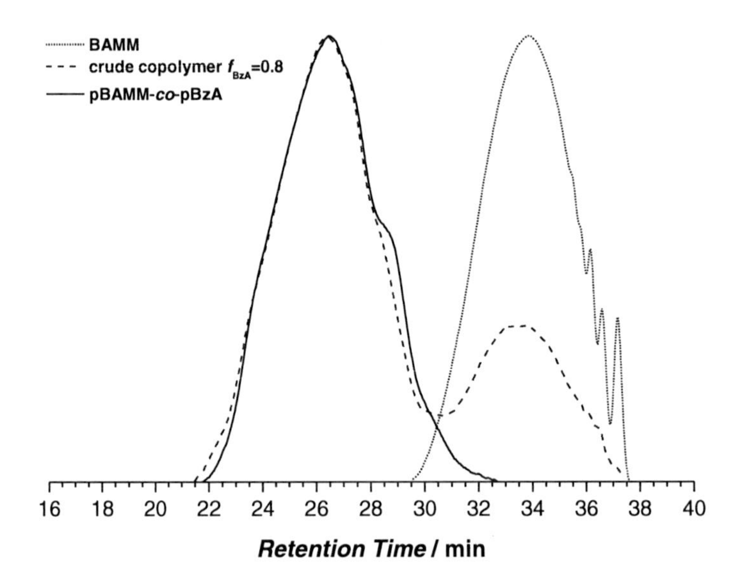

Figure 8-3. SEC elugrams of BAMM (dotted line), the crude product after the reaction (dashed line) and the copolymer (f_{BzA} = 0.8; straight line) after threefold precipitation in an excess of cold methanol. Copolymerization conditions: BAMM, BzA, toluene, 1×10^{-1} mol·L^{-1} AIBN, 60 °C, reaction time 1-3 h (depending on f_{BzA}) up to approximately 40 % conversion.

The crude copolymer (dashed line) obtained from f_{BzA} = 0.8 displays a bimodal distribution (see Figure 8-3). Residual macromonomer (pure BAMM represented by the dotted line) appears at higher retention times, yet with reduced intensity. The residual BAMM co-monomer is detectable in the crude pBAMM-*co*-pBzA due to incomplete conversion and needs to be eliminated. Threefold precipitation of the crude product removes the residual macromonomer at higher retention times. Thus, the copolymer pBAMM-*co*-pBzA (represented by the straight line) exclusively shows a distribution at lower retention times, i.e., at higher molecular weights. The intensity of the residual macromonomer in the SEC elugrams of the crude copolymer (dashed line) increases with a decrease in f_{BzA}. This is shown in Figure 8-4 representing the SEC elugrams for a co-monomer feed $f_{BzA} = 0.8$ (top) and $f_{BzA} = 0.1$ (bottom).

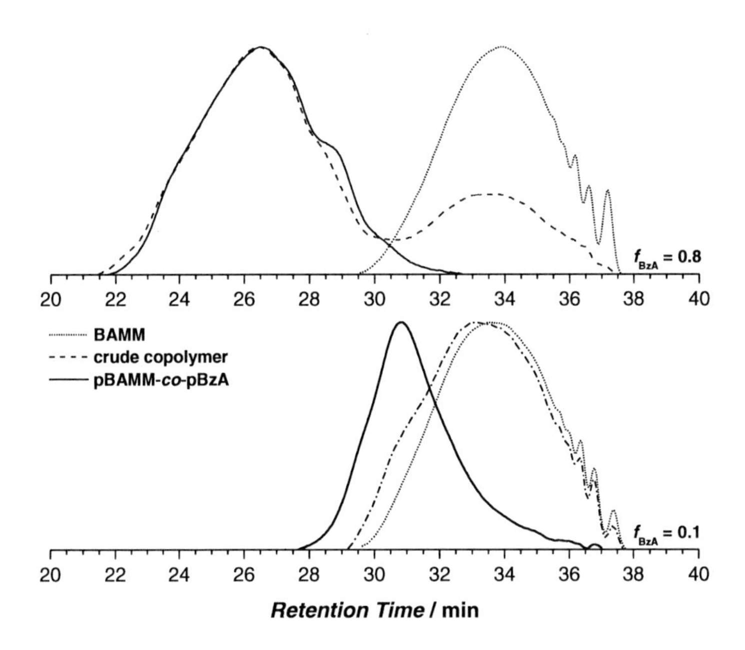

Figure 8-4. SEC elugrams of BAMM, the crude product after the reaction and the copolymer $f_{BzA} = 0.8$ (top) and the copolymer $f_{BzA} = 0.1$ (bottom), respectively, after three fold precipitation in methanol.

A further precipitation is required for each sample to remove residual co-monomer resulting in a pure copolymer. Evidence for the complete removal of the macromonomer by precipitation was obtained from reference SEC measurements of polymer blends. To this end, benzyl acrylate homopolymer (pBzA) has been synthesized via FRP with a molecular weight of $M_n = 59500\ g{\cdot}mol^{-1}$ and mixed with BAMM in a certain ratio, i.e., 9 : 1 pBzA : BAMM. The resulting polymer blend BAMM-*blend*-pBzA has been subjected to SEC. Even in a blend with 10 % of BAMM (see Figure 8-5) the macromonomer is detectable at higher retention times. Since – after the precipitation step – the SEC elugrams show no residual distribution on the low molecular weight side, it is certain that no or if any only minor amounts of macromonomer (<< 10 %) remains within the copolymer sample. The copolymers were analyzed via SEC using Mark-Houwink parameters for pBA

($K = 12.2 \times 10^{-5}$ dL·g^{-1}, $\alpha = 0.70$).[290] The resulting values for molecular weight and polydispersity are shown in Figure 8-6.

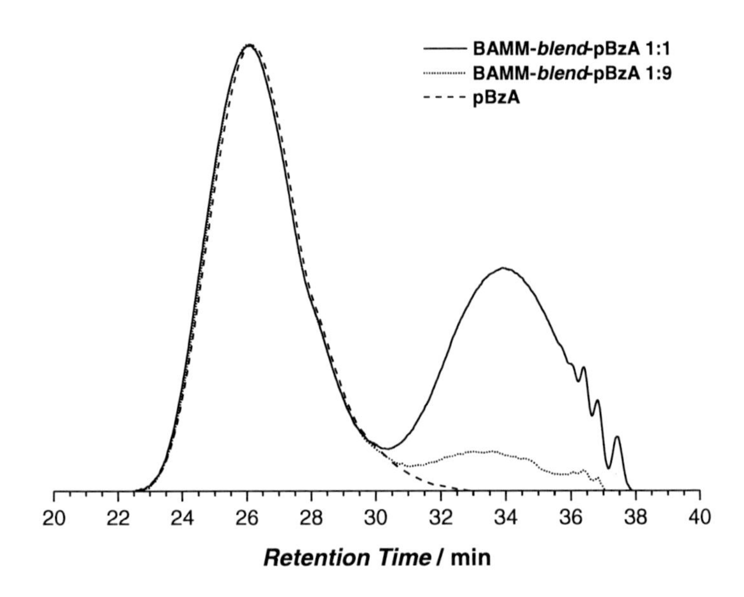

Figure 8-5. SEC elugrams of pBzA and BAMM mixtures at 1:1 and 1:9 ratios, respectively.

The lower part of Figure 8-6 depicts the dependence of the number-average molecular weights (represented by the dashed line) on the mole fraction of f_{BzA} in the initial co-monomer feed. With increasing mole fraction of benzyl acrylate, the molecular weight of the resulting copolymer increases, i.e., the copolymer obtained at $f_{BzA} = 0.9$ is close to a number-average molecular weight of 77000 g·mol^{-1} as the upper limit, whereas the copolymer generated at $f_{BzA} = 0.1$ has a number-average molecular weight of 8000 g·mol^{-1} as lower limit. The polydispersity of the copolymers (upper part of Figure 8-6) lies in the range between 1.30 to 2.12 and increases with increasing f_{BzA}. The number-average molecular weights and polydispersities (*PDI*s) of the copolymers have been determined using the Mark-Houwink parameters of pBA homopolymers; thus, the values obtained via SEC analysis should be regarded as estimated values. The low *PDI*s appear as a consequence of the purification step, since low molecular weight copolymeric species are discriminated and removed simultaneously with the residual macromonomer. Inspection of Figure 8-6 suggests that the propagation rate and the probability of insertion of the macromonomer is significantly lower compared to that of benzyl acrylate. Such an observation is based on the resulting molecular weights at different mole fractions of co-monomer BzA. The higher the co-monomer feed of BzA in the co-monomer mixture the higher the achievable molecular weight. With a low f_{BzA} value only small amounts of BzA are present at the beginning of the polymerization which leads to lower molecular weights due to a potentially lower propagation rate of the macromonomer. In a subsequent step, the copolymer composition is determined.

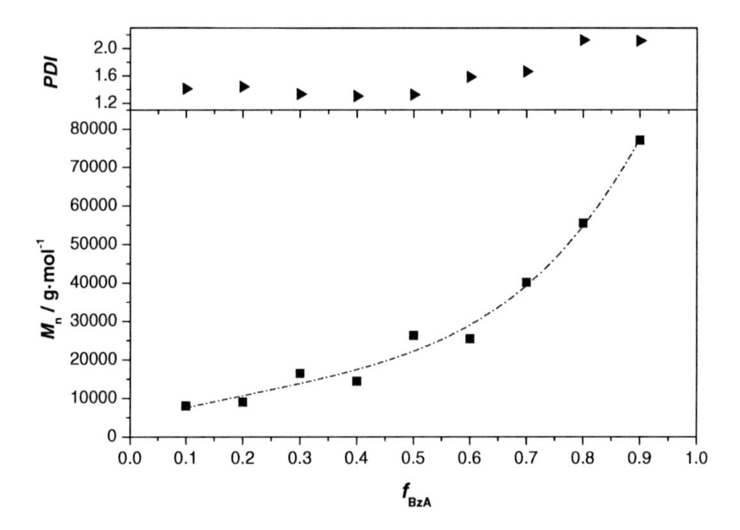

Figure 8-6. M_n (bottom) and *PDI* (top) as a function of the co-monomer feed of benzyl acrylate f_{BzA} for the synthesized statistical copolymers pBAMM-*co*-pBzA. Copolymerization conditions: BAMM, BzA, toluene, 1×10^{-1} mol·L^{-1} AIBN, 60 °C, reaction time 1-3 h (depending on f_{BzA}) up to approximately 40 % conversion.

8.5.2.2 Determination of Copolymer Composition

The determination of the copolymer composition for statistical copolymers obtained at variable initial BzA feed ratios, f_{BzA}, can open an avenue for a deeper understanding of the copolymerization process. For the determination of the copolymer composition, ^{1}H-NMR spectra were recorded for polymers generated from experiments at variable mole fractions of the co-monomer BzA. A typical ^{1}H-NMR spectrum of a copolymer pBAMM-*co*-pBzA is depicted in Figure 8-7.

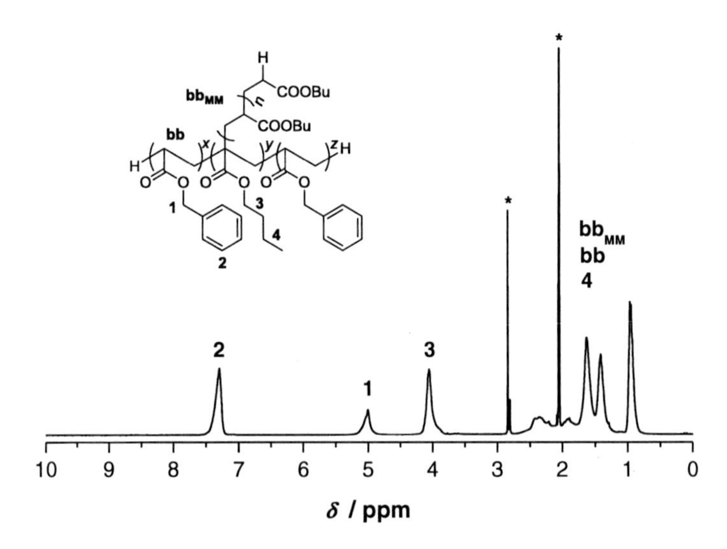

Figure 8-7. Typical 400 MHz ^{1}H-NMR spectrum of pBAMM-*co*-pBzA (f_{BzA} = 0.8). The asterisks represent acetone and water in acetone, respectively. The structural formula with the detected resonances is inserted. Copolymerization conditions: BAMM, BzA, toluene, 1×10^{-1} mol·L^{-1} AIBN, 60 °C, 3 h reaction time up to approximately 40 % conversion.

The assignments of the detected resonances are embedded in the structural formula within Figure 8-7. Benzyl acrylate was chosen as a co-monomer due to the presence of an aromatic phenyl ring, which causes a broadened signal H_{ar} (2) at ≈ 7.3 ppm, representing the aromatic protons in the backbone benzyl ring, separated from other signals and the signal of CH_2 (1) at ≈ 5.01 ppm for the methylene protons next to the oxygen of the acrylic side chain. As a counterpart representative of the fraction of BA in the statistic copolymer, the signal for the methylene protons next to the oxygen (3) at ≈ 4.05 ppm in the side chain of the macromonomer was selected. The resonances in the region below 3 ppm are related to the protons of the polymer backbone, both benzyl acrylate and *n*-butyl acrylate macromonomer, and the alkyl chain of the *n*-butyl side chain of the macromonomer; these signals are not relevant for the copolymer composition determination and are thus not considered further. NMR spectra have been recorded for each prepared copolymer and Figure 8-8 depicts the relevant NMR spectral region of the sample obtained for f_{BzA} = 0.9 (a) and f_{BzA} = 0.1 (b) as the upper and lower limits of the feed composition variation.

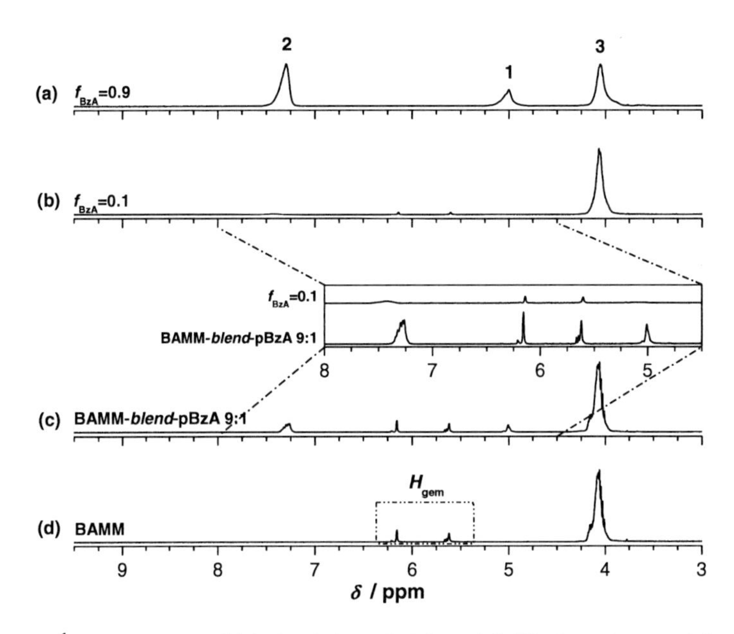

Figure 8-8. ^{1}H-NMR spectra of (a) pBAMM-*co*-pBzA f_{BzA} = 0.9, (b) pBAMM-*co*-pBzA f_{BzA} = 0.1, (c) polymer blend BAMM-*blend*-pBzA (9:1), (d) BAMM. Copolymerization conditions: BAMM, BzA, toluene, 1 × 10^{-1} mol·L^{-1} AIBN, 60 °C, 1h (f_{BzA} = 0.9) and 3h (f_{BzA} = 0.1) reaction time up to approximately 40 % conversion. The inserted box displays the comparison of the copolymer synthesized at f_{BzA} = 0.1 and the BAMM-*blend*-pBzA (9:1 wt%); note that the resonance CH_2 (3) of the macromonomer has been normalized for the spectra (2) and (3) in the inserted box; the polymeric signals H_{ar} (2) and CH_2 (1) of the BzA in the copolymer are less intense compared to the polymer blend due to different number-average molecular weights (8000 g·mol^{-1} for the copolymer and 63500 g·mol^{-1} for the pBzA of the polymer blend).

To evidence that the macromonomer has indeed undergone copolymerization, the following strategy is adapted: A ^{1}H-NMR spectrum of a blend of 90 wt% BAMM and 10 wt% pBzA is recorded and compared in the vinylic region to a copolymerization run of an initial ratio f_{BzA} = 0.1 and BAMM, respectively. If BAMM is indeed incorporated in a copolymer, a significant reduction in the geminal proton signal H_{gem} should be observed. Ideally, no geminal proton signal should remain in the copolymerization experiment. Spectrum (a) (Figure 8-8) depicts the case where f_{BzA} = 0.9. As it can be observed from spectrum (b), which represents f_{BzA} = 0.1, little residual macromonomer remains in the copolymer, as evidenced by the appearance of the resonances H_{gem} at ≈ 5.62 ppm and ≈ 6.16 ppm for the geminal substituted olefine (pure BAMM is shown in spectrum (d)), featuring an f_{BzA} fraction of 0.3. Compared to the above described SEC measurements, the NMR characterization shows a higher sensitivity due to the fact that residual BAMM is visible in the spectra having fractions below $f_{BzA} \leq 0.3$. With the decrease of the co-monomer f_{BzA} the separation of the copolymer and residual BAMM by precipitation in cold methanol is becoming increasingly difficult. Nonetheless, copolymer is detectable as the major product, evidenced by the appearance of the signals CH_2 (1) and H_{ar} (2) of the benzyl acrylate co-monomer, respectively. The ^{1}H-NMR spectrum of a BAMM-*blend*-pBzA (9:1) (c) evidences a substantial decrease of the vinylic proton resonances H_{gem} after the polymerization process in spectrum (b) since the intensity is much lower (≈72 %) compared to the intensity in the spectrum of the polymer blend (c) (see inserted box of Figure 8-8). For the comparison of the spectrum of $f_{BzA} = 0.1$ (b) and the spectrum of the polymer blend (c) resonance CH_2 (3) of the macromonomer at ≈ 4.05 ppm has been normalized, hence the resonances H_{gem} in both spectra have been collated referring to their intensity. In addition, the decrease of the geminal resonances H_{gem} and the integral of CH_2 (3) confirms the incorporation of the co-monomer BAMM in the copolymer. In the pure BAMM spectrum (d), the integrals of the resonance H_{gem} – compared to the integrals of the resonance CH_2 next to the oxygen in the acrylate moiety – are represented to be $2\,H : 4\,H \times [M_n(\text{BAMM})/M(\text{BA})]$. Taking into account that the resonance CH_2 (3) of the copolymer potentially represents both residual BAMM and BAMM grafted in the copolymer, the integral needs to be higher than the integral of the pure BAMM, which is associated with the measured copolymers.

For the relevant area from 3 to 9 ppm, Figure 8-9 shows all ^{1}H-NMR spectra from a mole fraction of 0.1 to 0.9 BzA in the co-monomer mixture in a three dimensional plot. In Figure **8-9**, the resonance of the methylene protons of the macromonomeric part CH_2 (3) have been normalized according to the co-monomer feed f_{BzA} of the copolymerization mixture, i.e., the intensity of resonance (3) was set to 0.9 for the sample $f_{BzA} = 0.9$, the intensity of resonance (3) was set to 0.8 for the sample $f_{BzA} = 0.8$ etc. At high amounts of BzA, starting with $f_{BzA} = 0.9$, the intensity of both resonance (1) and (2) decrease with decreasing mole fraction of BzA. This implies that the resonances resulting from benzyl acrylate reduce with less available co-monomer. A deeper understanding of the copolymers requires a more detailed consideration not only in terms of a qualitative observation of the signals but also the quantification of the integrals resulting from the relevant resonances, which characterize the copolymer.

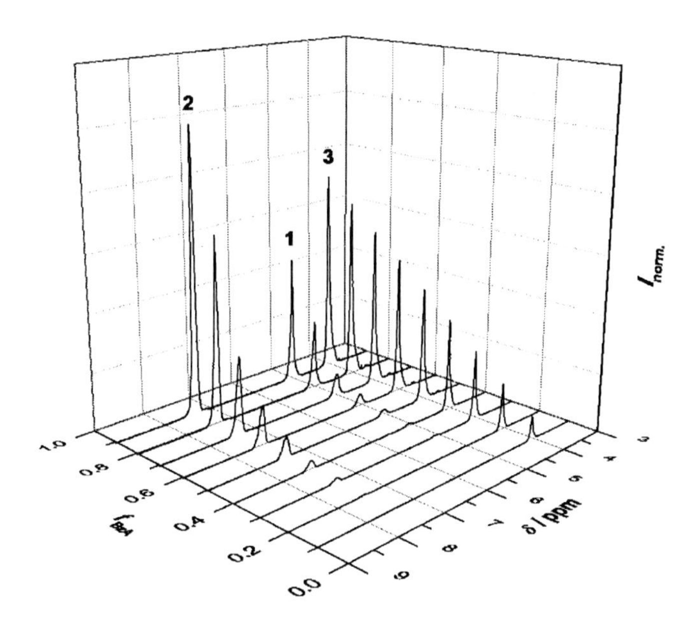

Figure 8-9. 3D ^{1}H-NMR plot of pBAMM-*co*-pBzA copolymers synthesized at variable mole fractions f_{BzA}. Shown is the area from 3 to 9 ppm highlighting on the H_{ar} resonance of BzA (1), the CH_2 resonance of BzA (2) and CH_2 (3) resonance of BAMM.

For the determination of the copolymer composition F_{BzA}, both signals H_{ar} (2) and CH_2 (1) (BzA) were integrated and compared to the integral of signal CH_2 (3) associated with the amount of macromonomeric repeating units. The mole fraction F_{BzA} in the copolymer is given by the equations (8-9) and (8-10) below:

$$F(\mathrm{BzA}) = \frac{I(H_{\mathrm{ar\ BzA}})}{\dfrac{2\, I(CH_{2\ \mathrm{BA}}) \times M(\mathrm{BA})}{M_n(\mathrm{BAMM})} + I(H_{\mathrm{ar\ BzA}})} \tag{8-9}$$

for the aromatic protons H_{ar} of the benzyl acrylate (2), or

$$F(\mathrm{BzA}) = \frac{I(CH_{2\ \mathrm{BzA}})}{\dfrac{I(CH_{2\ \mathrm{BA}}) \times M(\mathrm{BA})}{M_n(\mathrm{BAMM})} + I(CH_{2\ \mathrm{BzA}})} \tag{8-10}$$

for the methylene protons CH_2 (1) of the acrylic side chain of the benzyl acrylate, respectively. In other words, the BAMM is treated as a monomer with a molecular weight close to $M_n = 2300\ \mathrm{g{\cdot}mol^{-1}}$, i.e., a brush generating unit for the entire copolymer, whereas the benzyl acrylate generates a repeat unit of $M = 162.19\ g\cdot mol^{-1}$. The polymerizable group of the BAMM represents the geminal 1,1'-disubstituted double bond as a pseudo methacrylate, where the substituents are a *n*-butyl ester and the polymer backbone of the macromonomer. However, due to the fact that the macromonomer is present in the copolymer sample, the resonance CH_2 (3) displays the amount of macromonomeric repeating units in the copolymer as well as the residual macromonomer. Thus, the integral of resonance (3) has been reduced by the amount of residual macromonomer derived from the occurrence of the vinylic protons H_{gem}. Within the spectra the error obtained from residual macromonomer in the copolymer samples $f_{BzA} \leq 0.3$ is insignificant as the amount of residual co-monomer reaches 1.64 % at maximum for $f_{BzA} = 0.1$ as the highest amount of utilized macromonomer. The resulting values for the mole fraction in the copolymer F_{BzA} are plotted against the initial mole fraction f_{BzA} of the co-monomer in the mixture. The resulting Mayo-Lewis Plot is shown in Figure 8-10.

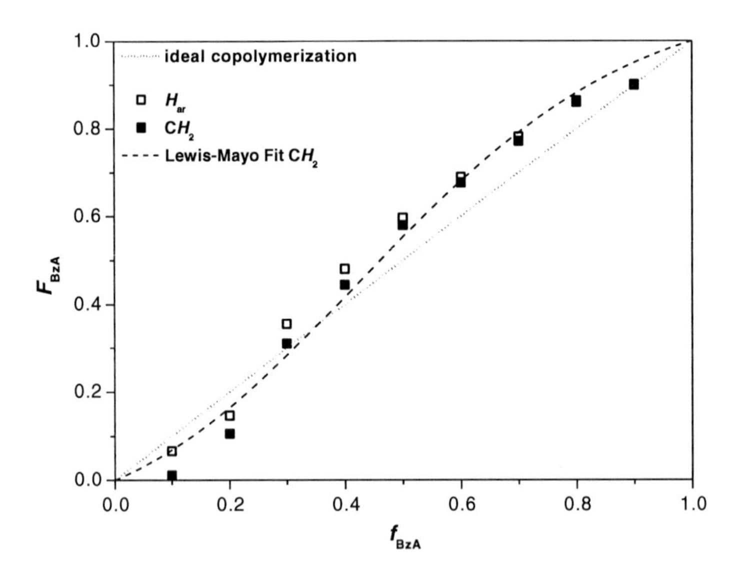

Figure 8-10. Mayo-Lewis plot for the entire copolymerization of BAMM with BzA in toluene with 1×10^{-1} mol·L^{-1} AIBN at 60 °C to approximately 40 % of conversion determined gravimetrically. Values are determined from NMR integrals CH_2 (3) vs. CH_2 (1) and H_{ar} (2), respectively. The derived reactivity ratios (see text) should be seen as estimated values.

The hollow squares in Figure 8-10 represent the values determined from the aromatic resonances H_{ar} (2) whereas the filled squares are derived from the integration procedure involving the methylene resonances CH_2 (1) of the benzyl acrylate co-monomer. It can be observed that the values F_{BzA} for H_{ar} are slightly higher compared to those derived from CH_2.

This is caused by residual toluene in the samples after high vacuum overlapping with resonances of the H_{ar} (2) signal of the benzyl acrylate. Consequently, the values obtained from the resonances CH_2 (1) represent the copolymerization behavior of both monomers more accurately. Further, the calculation of the specific integrals from the measured ^{1}H-NMR spectra – providing copolymer composition F_{BzA}– can be used for estimating the amount of grafted BAMM onto the copolymer backbone as a percentage of the overall molecular weight of the copolymer. The obtained number-average molecular weights for each sample from $f_{BzA} = 0.9$ to $f_{BzA} = 0.1$ have been employed for a detailed analysis. The percentage of the grafted BAMM on the polymer backbone starts at ≈ 0.78 % for $f_{BzA} = 0.9$ and increases nearly linear up to 13.6 % for $f_{BzA} = 0.3$. From $f_{BzA} < 0.3$, the amount of grafted BAMM increases rapidly up to 86.5 % BAMM on the polymer backbone for $f_{BzA} = 0.1$ (see Figure 8-11).

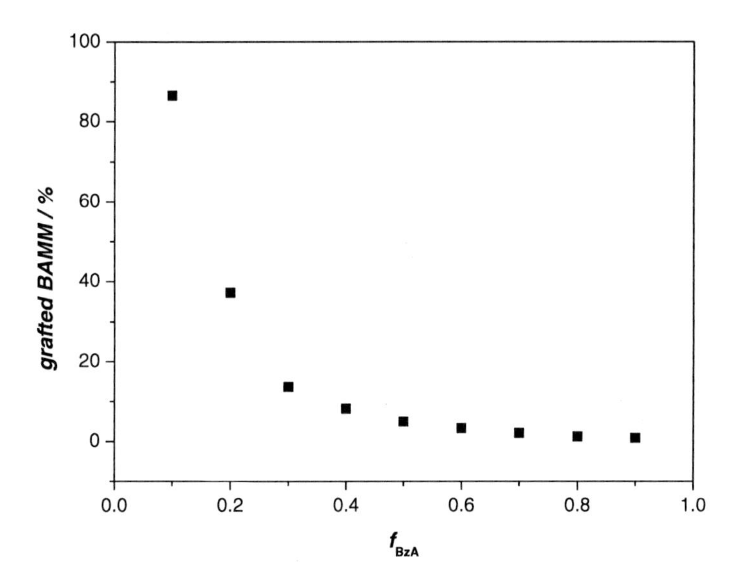

Figure 8-11. Percentage of grafted BAMM on the copolymer backbone calculated from number-average molecular weights M_n obtained from SEC.

The determination of the reactivity ratio of the copolymerization was accomplished by further calculations based on the Mayo-Lewis equation (8-7). The experimental data from various co-monomer feeds f_{BzA} were applied to a linear squares regression analysis to calculate reactivity ratios of the copolymer composition which generally requires low conversion experiments. However, the experiments were carried out to approximately 40 % conversion due to precipitation restrictions of the copolymers at lower conversions. Thus, the reactivity ratios are calculated from the experimental composition data for an approximate estimation of the macromonomer reactivity in a free radical copolymerization with an acrylate; they are thus estimates, as it is clear that in the case of 100 % conversion both values reach unity. However, experimental restrictions associated with polymer purification do not allow isolating copolymers at lower conversions. Fitting of the experimental data via the Mayo-Lewis equation (8-7) leads to the reactivity ratios of BAMM and BzA. The reactivity ratios originating from the NMR CH_2 (1) resonance (dashed line, Figure 8-10) and from the NMR H_{ar} (2) resonance are collated in Table 8-1.

The dotted line embedded in Figure 8-10 represents ideal copolymerization behavior with $r_{i,j} = 1$, meaning each monomer has the same addition probability to each propagation species, resulting in a statistical copolymer of a composition equal to the feed monomer ratio.

Table 8-1. Estimated *r*-values for the copolymerization process of BAMM with BzA in toluene.

	r_1 (r_{BzA})	r_2 (r_{BAMM})
	(≈ 40 % conversion)	
H_{ar} [a]	2.12	1.26
CH_2	2.46	1.79

[a] residual toluene in the samples overlapping with resonances of the H_{ar} (2) signal of the benzyl acrylate causes higher F_{BzA}; the values obtained from the resonances CH_2 (1) represent the copolymerization behavior of both monomers more accurately.

The integrals obtained from the aromatic resonances (2) are negligible, since aromatic protons of the residual toluene are counted additionally, leading to overestimated mole fractions in the copolymer feed. The resulting *r*-values describe the system in a reasonable fashion, yet it needs to be clearly noted that the ratios have been calculated based on copolymers synthesized at approximately 40 % conversion. Thus, the reactivity ratios represent an estimation; nevertheless, the values proof that the macromonomer synthesized via the high-temperature one-pot – one-step procedure undergoes copolymerization with small acrylate monomers. Interpretation of the calculated reactivity ratios – considering their estimated nature – leads to the following: $r_1 = 2.46$ representing the propagation tendency of benzyl acrylate. The radical character on a benzyl acrylate terminus of the copolymer prefers to propagate with the identical species, namely a benzyl acrylate monomer, whereas the same applies to the macromonomer, which propagates preferentially with itself resulting in $r_2 = 1.79$. When considering both reactivity ratios, the benzyl acrylate propagation step is favored. Such an observation may be associated with the steric demand of the macromonomer, especially if two BAMMs are placed next to each other within the copolymer. However, it is assumed that the lower molecular weight chains of the macromonomer (*PDI* = 1.90) potentially have a higher propagation probability due to decreased steric hindrance. In contrast, benzyl acrylate is sterically unambitious and shows a higher propagation rate related to the macromonomer. The understanding on a molecular level is correlated with the molecular weights (see Figure 8-6) achieved from the copolymerization process. The decrease of molecular weight with the decrease of f_{BzA} generates a low molecular copolymer due to lower propagation of the in excess present macromonomer and the low availability of the faster propagating benzyl acrylate in these samples. For $f_{BzA} > 0.35$, the Mayo-Lewis fit (dashed line) lies above ideal copolymerization (dotted line) and below the ideal for lower f_{BzA} content. The azeotropic point of the copolymer composition, where f_{BzA} equals F_{BzA}, is reached at approximately $f_{BzA} = 0.35$. The copolymerization of high-temperature acrylate polymerization macromonomers thus leads to statistic copolymers with benzyl acrylate monomers in the polymer backbone as well as grafted BAMM units, depending on the amount of macromonomeric repeat units in the copolymer.

It is additionally important to note that even though the (benzyl acrylate) radical chain terminus has a high propensity to add a monomeric acrylate monomer, addition of a BAMM unit is not much slower and the individual rate coefficient of addition is within a factor of 2-3 of the homo-propagation rate of the acrylate. For the modelling of high-temperature acrylate polymerization, this rate coefficient was to date not known with sufficient certainty and while simulation results indicated fairly high addition rates of macromonomer units to growing chains,[64,291] no direct experimental data were available so far. The agreement between the copolymerization parameter of 2.46 determined herein and the value obtained by Hutchinson and colleagues for high-temperature homopolymerization of BA (1.82)[291] is – considering the indirect approach to determine the value via kinetic simulations – remarkable.

The copolymerization behavior of structurally very similar macromonomers and the estimation of their reactivity ratios have been studied before. Among others, Hirano *et al.* report in 2003 the free radical copolymerization of unsaturated oligomers (n = 2-4) of methyl acrylate with cyclohexyl acrylate.[292] They obtained reactivity ratios of the methyl acrylate tetramer of $r_1 = 0.54$ and $r_2 = 0.96$.[292] These dimer, trimers and tetramers of methyl acrylate featured the same olefinic terminus and should thus be expected to behave similar to the macromonomers under investigation in the present study. It could, however, be expected that the steric demand of these significantly smaller macromonomers plays a significant role and may be responsible for the different reactivity (e.g., faster homo-propagation of the macromonomers and increased propensity to add to an acrylate chain terminus). In the light of this, it should be noted that the values determined herein represent an average over the whole distribution of macromonomer and that a chain length dependency of the *r*-values might exist.

In addition, Ohno *et al.* reported in 2006 that the reactivity ratios of macromonomers synthesized via post-modification of ATRP polymers of BA with BA are close to unity.[287] These parameters were, however, calculated from high conversion data; in this case it is not surprising that the reactivity ratios are close to unity since the parameters must reach unity for 100 % conversion.

The above interpretation of the data is based on the treatment of the macromonomer as a co-monomer, a view that underpins the calculations above and the Mayo-Lewis diagram. However, for a more detailed understanding, the copolymer needs to be analyzed in terms of its two monomeric constituents, *n*-butyl acrylate and benzyl acrylate. For each macromonomeric repeat unit in the final copolymer an average of ≈ 18 BA units ($2300\ g{\cdot}mol^{-1}/128.17\ g{\cdot}mol^{-1}$) are counted. Thus, the amount of BA is 18 times higher than the amount of benzyl acrylate related to the monomeric species. This fact is clarified in Figure 8-12, where the amount of BA and BzA units, respectively, is plotted in contrast to Figure 8-11, where the macromonomer BAMM is considered as one constituent with approximately 18 BA repeat units.

The amount of BA in the copolymer is represented by the filled squares, whereas the amount of BzA is represented by the hollow squares, respectively. A rapid increase from $f_{BzA} = 1$, which would yield a homopolymer with $F_{BA} = 0$, to $f_{BzA} = 0.9$ is visible in Figure 8-12 and reaches a BA copolymer composition of $F_{BA} = 0.64$. At a co-monomer feed of $f_{BzA} = 0.1$ the percentage of incorporated benzyl acrylate is (almost) quantitative. The above analysis of the copolymer composition and its relation to the co-monomer feed, in terms of the molecular building blocks, forms the basis for understanding the chromatographic results of the copolymer at critical conditions of pBA.

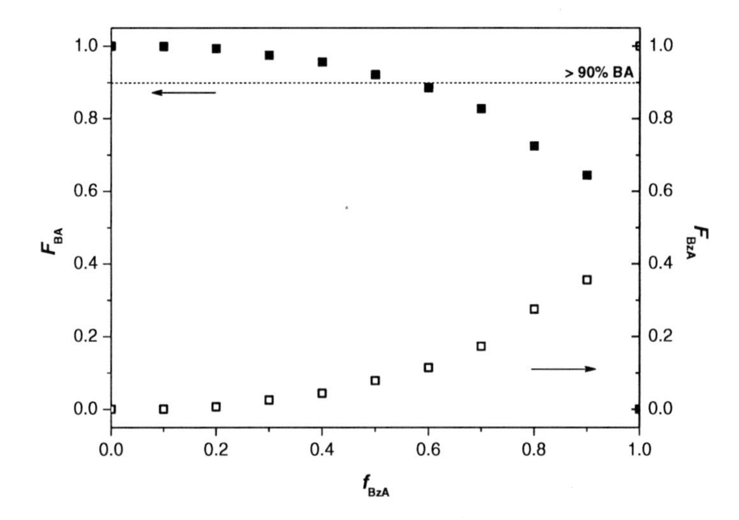

Figure 8-12. Visualization of the co-monomer feed of BA (filled squares) and BzA units (hollow squares), respectively, in the copolymer against the co-monomer feed of BzA in the co-monomer mixture. The squares in the region above the dotted line represent ≥ 90 % BA in the copolymer.

Chromatographic analysis of the statistical copolymers can offer a detailed understanding of the samples' composition. In principle, employing LCCC enables to assess the main part of the sample: the quantity of BAMM and BzA embedded in the copolymers. Such information can be more detailed than quantification via NMR as described above. When operated in 2D-mode (thus in conjunction with size exclusion chromatography), separation for chemical composition as well as molecular weight is accessible, thus allowing to differentiate between copolymer of BzA and BAMM and the respective blend of homopolymers.

8.5.2.3 Characterization of the Copolymers by LCCC

Critical conditions for pBA were determined and utilized to perform the analysis of the copolymers on a LCCC system. When measured on a high pressure liquid chromatography (HPLC) system at 100 % of a polymer soluble solvent – THF for pBA – the copolymer samples elute according to their molecular weight in a narrow elution time window (see Figure 8-13 dashed line), whereas macromolecules can potentially be separated under critical conditions according to their composition (Figure 8-13 straight line for a copolymer sample and the dotted line for pure macromonomer).

For that reason, pBA standards with different molecular weights have been synthesized via reversible addition fragmentation transfer (RAFT) polymerization.[293-294] Under critical conditions of adsorption, the entropic and enthalpic effects between the polymer samples and the packing column are compensated. At the critical conditions of a specific polymer, the macromolecules elute at the same retention volume independent of their molar mass. This implies that the first dimension (LCCC) separates no longer by size (SEC mode) but according to the chemical functionality. Critical conditions of pBA on a normal phase column are realized at a solvent mixture of THF and *n*-hexane with a ratio of 28.7 to 71.3 (v/v) at 35 °C. The pBA samples elute at a retention volume of around 3.38 mL regardless of their molecular weight. For a detailed graphical description of the critical conditions see Chapter 4.

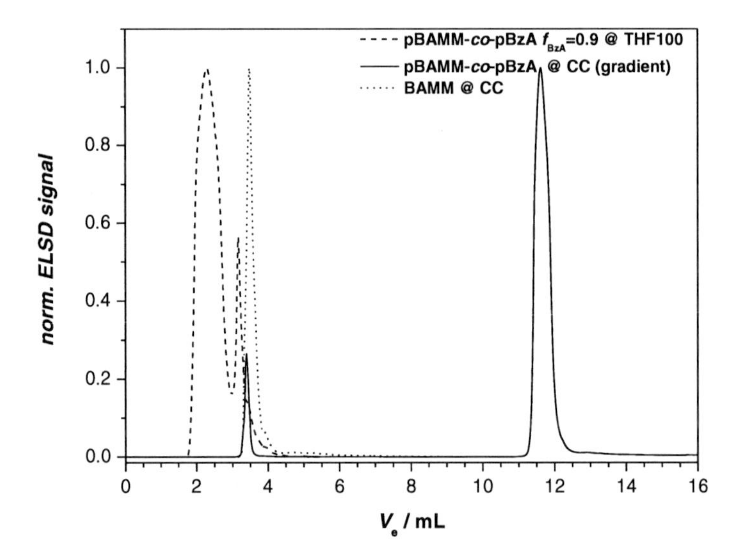

Figure 8-13. Comparison of the elugrams of pBAMM-*co*-pBzA with a mole fraction of 0.9 of benzyl acrylate at critical conditions (applied gradient) and 100 % THF; the elugram of BAMM at critical conditions is additionally shown.

LCCC represents a useful tool for block copolymer characterization since the copolymers are separated according to their chemical functionality and the segment of the critical conditions gets invisible.[112,118,121] The LCCC measurement of the copolymers pBAMM-*co*-pBzA is applied to separate the various samples according to the incorporated amount of BzA in the polymer backbone analogous to the studies of Falkenhagen *et al.*[112] However, the characterization of the statistic copolymers shows a different chromatographic behavior due to different macromolecular composition compared to block copolymers. For the measurement of the copolymers pBAMM-*co*-pBzA a solvent gradient was applied to reduce the measuring time (for 100 % THF elugram see Figure 8-13). After 13 minutes of critical conditions the fraction of THF was allowed to reach 100 % (over a period of 10 minutes) a condition that was applied for further 12 minutes. The utilized gradient is depicted in Figure 8-14 (top). After the elution volume of around 10 to 13 mL the actual copolymer is detectable. While separation of the polymers under critical conditions is possible, it should be noted that the elution volume for complete elution of the injected sample is fairly high. Thus, in order to apply size exclusion chromatography and to also assess the chain length of the eluted species (which requires a strong reduction in the flow rate of the LCCC dimension) unreasonably long retention times would be required and thus the analysis of the present copolymers had to be restricted to the HPLC mode.

Figure 8-14 shows LCCC elugrams of both the selected copolymers with a co-monomer feed f_{BzA} of 0.9 to 0.7 and 0.1 and pure BAMM under critical conditions of pBA. The upper part of Figure 8-14 depicts the above described gradient.

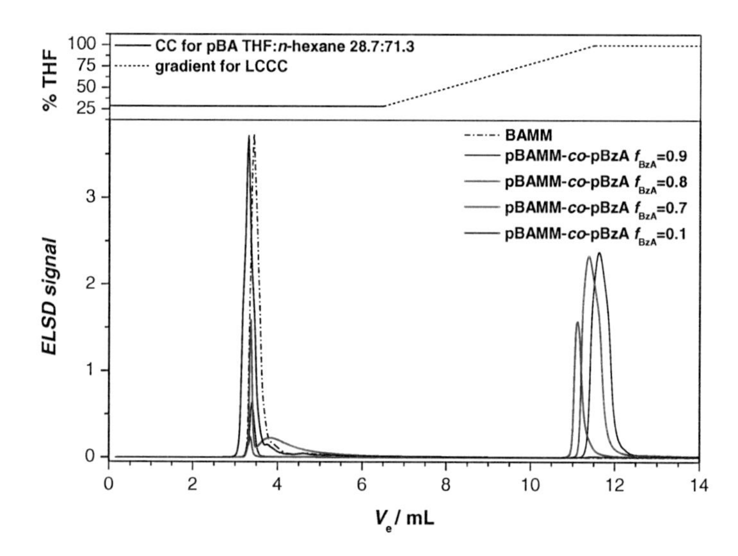

Figure 8-14. LCCC elugrams of different copolymer samples prepared at mole fractions 0.9 to 0.7 and 0.1 of BzA at critical conditions of pBA; the above insert shows the solvent gradient applied for the measurements on a normal phase column at 35 °C.

The copolymer samples appear to show residual macromonomer on the LCCC indicated by a distribution at the retention volume of pBA (V_e = 3.23 mL to 3.81 mL) even for higher f_{BzA}. The samples elute in adsoption mode within the gradient and pure THF regime (V_e = 10.9 mL to 12.4 mL), which means that the copolymer produced at f_{BzA} = 0.9 (highest molecular weight) is detectable at higher retention volumes (V_e = 11.2 mL to 12.4 mL) than the copolymers prepared at f_{BzA} = 0.7 (V_e = 11.1 mL to 12.3 mL) and f_{BzA} = 0.8 (V_e = 10.9 mL to 11.8 mL). The copolymers elute depending on their molecular weight with increasing retention volume in ascending order. The elugrams appear to indicate that the amount of residual macromonomer (V_e = 3.23 mL to 3.81 mL) increases with the decrease of the mole fraction of BzA in the reaction mixture since the peak intensity increases with decreasing f_{BzA}. However, considering the amount of BA monomer units in the copolymers the signals at V_e = 3.23 mL to 3.81 mL can readily be understood as minor amounts of residual macromonomer, but even more important, additionally as copolymeric species with a considerable proportion of BA. As Figure 8-12 indicates, the amount of BA – for which critical conditions were applied – increases rapidly with decreasing amounts of BzA as each macromonomer – when incorporated in the copolymer – contributes ≈ 18 BA units. Even for the copolymer prepared at f_{BzA} = 0.9 the amount of BA lies above 60 % and with the continuous increase of BAMM in the co-monomer feed the incorporated BA in the copolymer reaches > 90 % for the copolymers synthesized at $f_{BzA} \leq 0.6$. Further, the amount of grafted BAMM on the polymer backbone increases for decreasing f_{BzA} – up to 86.5 % BAMM on the polymer backbone for f_{BzA} = 0.1 (see Figure 8-11) – which indicates an excess of BA compared to BzA in the copolymer. In general, the analysis of a statistical copolymer is not necessarily depending on the endgroup availability. In the case of a block copolymer, the elution behavior can easily be understood since block B represents a chain extension of the block A and the samples elute under critical conditions of A according to the length of block B. In the case of a statistic copolymer, as considered in the current study, A und B units are distributed over the entire copolymer. Since the mole fraction of BA in the copolymer is significantly higher compared to BzA (see Figure 8-12) and approximates 90 % already at $f_{BzA} \leq 0.6$, the copolymers appear – at a certain ratio, which could be considered as a turning point – at the elution volume of pure pBA homopolymer independent of their molecular weight on the LCCC. Thus, the high fraction of BA (> 90 % for samples prepared at $f_{BzA} \leq 0.6$) makes the copolymer chromatographically indistinguishable from the residual pure macromonomer. Thus, the copolymers could nearly be seen as a pBA homopolymer with small defects induced by the incorporation of BzA in the polymer backbone. For $f_{BzA} \leq 0.6$ only one trace is detectable at V_e = 3.23 mL to 3.81 mL based on the previous explanation. For the sample f_{BzA} = 0.7 the turning point in LCCC behavior is clearly visible, displayed by the appearance of two distributions, one at higher elution volume (higher amount of F_{BzA}) and a second broadened trace in the BA area representing copolymers with small BzA repeat units as defects. In Figure 8-14, the copolymer sample resulting from f_{BzA} = 0.1 is shown as an example which contains F_{BzA} < 0.01 and elutes at the critical elution volume of pBA.

Since pBzA homopolymer could potentially be formed during the copolymerization process as a side product it is necessary to eliminate this doubt. Therefore pBzA homopolymer has been synthesized both via free radical and controlled radical polymerization (RAFT). However, the insolubility of pBzA in the critical solvent mixture THF : *n*-hexane of the LCCC measurement removes pBzA prior to the LCCC measurements. In conclusion, due to the insolubility of the homopolymer pBzA, the LCCC chromatograms in Figure 8-14 show the copolymeric species pBAMM-*co*-pBzA.

8.5.3 Conclusions

The *n*-butyl acrylate macromonomers synthesized via the high-temperature acrylate one-pot – one-step polymerization have been successfully copolymerized with benzyl acrylate as a representative acrylate monomer. Free radical copolymerization has been carried out for various co-monomer feeds. The resulting statistic copolymers featuring pendant side chains show a wide range in molecular weight depending on the initial mole fraction of the co-monomer. The achievable molecular weight of the synthesized pBAMM-*co*-pBzA copolymers lies between 8000 $g{\cdot}mol^{-1}$ and 77000 $g{\cdot}mol^{-1}$ with a polydispersity of 1.30 to 2.12.

The copolymer samples pBAMM-*co*-pBzA have been subjected to several characterization methods including SEC, NMR spectroscopy and LCCC at the critical conditions of pBA to obtain a detailed understanding of the generated structures. The integration of specific NMR resonances of the copolymer has been carried out. Further, composition analysis via the (*terminal model*) Mayo-Lewis equation provides estimates for the reactivity ratios of BAMM and BzA as co-monomers. As the copolymerization has been carried out up to 40 % conversion, the obtained reactivity ratio can be seen as estimates for the copolymerization. The reactivity ratios $r_{BzA} = 2.46$ and $r_{BAMM} = 1.79$ indicate a copolymer composition of $F_{BA} < 0.65$ for the copolymers derived from $f_{BzA} > 0.9$ up to a copolymer composition of $F_{BA} > 0.9$ for copolymers with a co-monomer feed of $f_{BzA} < 0.6$. Generally, the obtained reactivity ratio is in overall good agreement with literature data on macromonomer addition to growing acrylate chains based on kinetic simulations.

It has thus been established that macromonomers prepared via the facile high-temperature synthetic approach featuring a broad molecular weight distribution can indeed copolymerize, rendering the synthesis of variable grafting density chain structures (depending on the initial mole fraction of f_{BzA}) a valuable tool for graft copolymer design.

9

Transformation of Macromonomers into Ring-Opening Polymerization Macroinitiators

9.1 Introduction to Ring-Opening Polymerization (ROP)

The ring-opening polymerization (ROP) of cyclic monomers such as esters (e.g., lactones, dilactones), amides (e.g., lactams), ethers, acetals and siloxanes gives rise to condensation polymers. The mechanistic steps of ROP include initiation, propagation and termination. The polymerization of a specific monomer via ROP is influenced by both kinetic and thermodynamic factors.[295] The relative stability of the monomers compared to the resulting macromolecules is dictated by the ring strain of the monomers, thus the extent to which polymerization is thermodynamically favored varies according to the ring size, except for six-membered ring structures. The highly polarized functionality within the cyclic structure of six-membered rings allows for the ring-opening of the cycle by a potential electrophilic or nucleophilic initiator. Generally, ionic (e.g., anionic or cationic), coordination-insertion, enzymatic and radical mechanisms are responsible for ring-opening processes. The monomer type, catalytic and initiating system and the nature of the activated species for the ring-opening process influence the mechanism occurring during polymerization. Among these, the catalysis system is of critical importance. Several catalysts, for example metal, organo- and enzyme catalysts, are known to mediate ROP.[19] The living character of ring-opening polymerization is evidenced by a linear correlation of the number-average

Parts of this chapter were reproduced with permission of John Wiley & Sons.[14]

molecular weight to the monomer conversion. Preselected polymer molecular weights are readily generated with narrow polydispersities. The mechanistic details of the mentioned catalytic systems for ROP are described in the following section.

9.2 Catalysis in Ring-Opening Polymerization

9.2.1 Metal Catalyzed ROP

ROP catalyzed by metal complexes is widely employed. Typical catalytic systems are tin(II) octanoate ($Sn(Oct)_2$), zinc(II) lactate ($Zn(Lact)_2$) and aluminium isopropoxide ($Al(O\textit{i}Pr)_3$) (Figure 9-1). Among the mentioned catalysts, $Sn(Oct)_2$ is used for industrial processes as it features the highest catalytic activity in ROP, only requiring the presence of an alcohol as co-initiator.

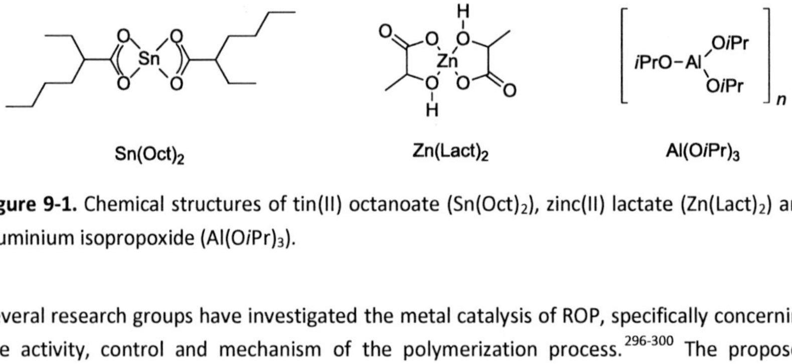

Figure 9-1. Chemical structures of tin(II) octanoate ($Sn(Oct)_2$), zinc(II) lactate ($Zn(Lact)_2$) and aluminium isopropoxide ($Al(O\textit{i}Pr)_3$).

Several research groups have investigated the metal catalysis of ROP, specifically concerning the activity, control and mechanism of the polymerization process.[296-300] The proposed mechanism follows a coordination-insertion process as shown in Scheme 9-1.[296-298,300-301]

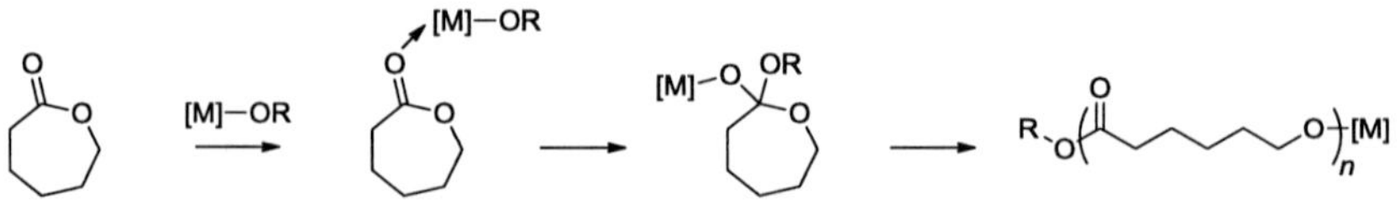

Scheme 9-1. Coordination-insertion mechanism of the metal catalyzed ring-opening polymerization of lactones (illustrated on ε-caprolactone).

The actual initiator of the ROP process, referred to as a metal alkoxide, is generated *in-situ* by the complexation of the metal to the alcohol co-initiator. Each alkoxide group on the metal is able to initiate a polymer chain, thus more than one polymer chain can be grown via the metal atom. For example, the propagation of two polymer chains is feasible for each $Sn(Oct)_2$ catalysis/initiation complex.

The metal-activated alkoxide complex, represented by [M]-OR, coordinates the carbonyl oxygen of the monomer (here ε-caprolactone) upon insertion into the monomer. Next, propagation occurs via subsequent incorporation (i.e., polymerization) of monomer until the monomer is consumed. In the last step the metal is removed from the terminus of the polymer chain by hydrolysis of the metal alkoxide bond.

Most often, relatively broad molecular weight distributions are observed for polymers synthesized via ROP. This observation can be explained by the potential side reactions that occur during the polymerization process.[302] The desired controlled polymerization process is often negatively affected by transesterification side reactions.[302-303] Two potential transesterification pathways are depicted in Scheme 9-2.

(a) [M]–O ... O–R ; R–O ... O–[M] ⟶ R–O(...O)₂–[M] + [M]–OR

(b) R–O ... O–[M] ⟶ ε-caprolactone + [M]–OR

Scheme 9-2. Intermolecular (a) and intramolecular (b) transesterifictaion reactions that are possible during the metal catalysis of ring-opening polymerizations. R represents either the initiation chain or the propagating chains.

The metal alkoxide coordinated to the monomer at the propagating polymer chain end is not just limited to further coordination with monomer existing in the system, but it is also capable of coordinating with propagating chains via the existing ester moieties on these linear polymer chains. For instance, intramolecular transesterification (Scheme 9-2 (a)) occurs between two (propagating) polymer chains. The metal terminus, therefore, coordinates to the ester carbonyl oxygen releasing a part of the polymer chain. Scheme 9-2 elucidates this transesterification mechanism for one repeat unit. In this case, the repeat unit of the obtained polymer chain is doubled while a metal alkoxide is released. Additionally, transesterification can occur within one propagating polymer chain; this process is referred to as intramolecular transfer. In contrast to intermolecular transesterification, intramolecular transesterification results in the formation of macrocycles. In addition, any other potential ester functionality in the system can, thus, be attacked by the propagating metal alkoxide complex.

9.2.2 Organo-Catalyzed ROP

Although, metal catalysis (as described above) is frequently used for ROP, the removal of residual metal catalyst in industrial applications is costly and often challenging due to the catalyst's toxicity. Given these drawbacks, ROP would be a more attractive process if it could be catalyzed by alternative catalytic systems. Fortunately, several metal-free catalysts, such as tertiary amines, phosphines and N-heterocyclic carbenes can be utilized for ROP. Herein, the superbasic guanidine 1,5,7-triazabicyclo[4.4.0]dec-5-ene (TBD) will be selected as a representative organo-catalytic system.

The catalytic activity of TBD in ROP was first observed in 2006.[304] The esterification, and thus chain propagation, was found to occur via the hydrogen bonding of the tertiary amine in the guanidine with the lactone.[304-305] The proposed mechanism for organo-catalyzed ROP is depicted in Scheme 9-3.

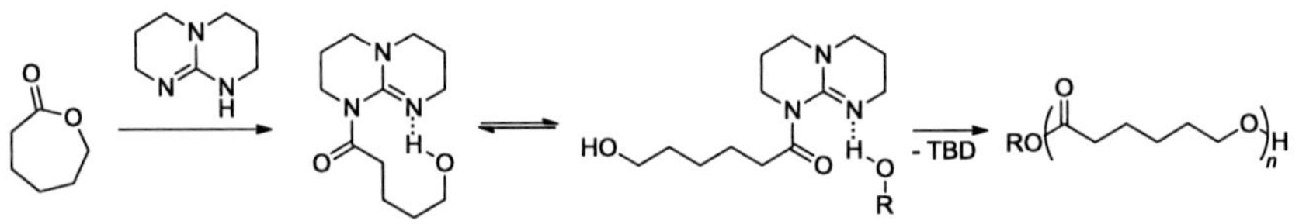

Scheme 9-3. Reaction mechanism for the organo-catalyzed (here, TBD) ring-opening polymerization. Dual activation of monomer and initiator occur.

TBD catalysis features the dual activation of both the monomer and the initiator.[306] Unfortunately, TBD is also known to catalyze transesterification reactions.[304-305,307-308] This side reaction can be observed for polymerizations performed up to high conversion, as is evidenced by the broad molecular weight distributions that are generated.

9.2.3 Enzyme-Mediated ROP

Enzymes are well known as *in vivo* catalysts for biochemical processes such as the formation of DNA, proteins and poly(hydroxyalkanoate)s; however, enzymes are also utilized *in vitro* for ROP catalysis.[309] Ester bond formation can be catalyzed with a lipase, a type of hydrolase, which features high chemoselectivity and regioselectivity. A typical lipase used in ROP is *candida antarctica* lipase B (CALB), which is commercially available in the form of macroporous crosslinked beads of poly(methyl methacrylate) on which the enzyme is adsorbed (trademarked as Novozym® 435). This enzyme demonstrates the highest activity in the ROP processes.[310-312] Enzyme mediated ROP was first reported in 1993 by two independent groups,[311,313] and since then the process has been investigated for its application to ε-caprolactone polymerization, by several research groups.[310,312,314-321] A general reaction mechanism was derived from experimental observations that progresses via an activated monomer mechanism as shown in Scheme 9-4.[317,320,322-323]

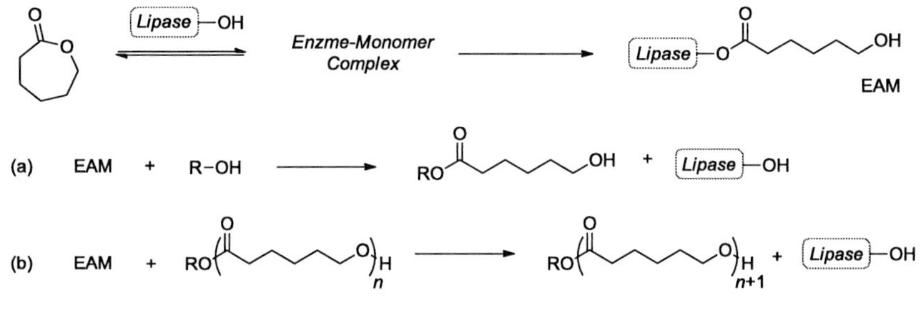

Scheme 9-4. Postulated reaction mechanism for enzyme mediated ring-opening polymerization of lactones (illustrated with ε-caprolactone as monomer), which proceeds via the formation of an enzyme activated monomer (EAM).[317,320,322-323]

The first step of this mechanism is the nucleophilic attack of the hydroxyl group within the lipase on to the monomer (e.g., ε-caprolactone) in Scheme 9-4, which generates the enzyme activated monomer (EAM) through an enzyme-monomer complex. Initiation of the polymerization (Scheme 9-4 (a)) occurs through the nucleophilic reaction of the added alcohol initiator with the EAM. During propagation (Scheme 9-4 (b)) the EAM undergoes nucleophilic attack by the terminal OH. This results in the release of the final polymer and thus the regeneration of the enzyme. The catalytic triade within the Novozym® 435 catalyst consists of the amino acids serine, histidine and asparagine.[295]

Analogous to the previously mentioned catalytic systems, transesterification can be observed in enzyme catalyzed ROP due to the presence of ester moieties in both the monomer and the propagating chain.[324-325] The extent of macrocyclization during the polymerization process is influenced by the reaction conditions including reagent concentrations (catalyst and initiator), solvent medium and temperature. However, the transesterification side reactions can be minimized for polymerizations performed in bulk and at low conversions.[326]

9.3 Ring-Opening Polymerization of Macroinitiators from High-Temperature Macromonomers

Recent developments in the area of industrial polymers focus on their renewability, biocompatibility and biodegradability as important aspects for polymer design and the choice of building blocks.[327-329] Aliphatic polyesters based on, i.e., lactide, glycolide or lactones combine the desired properties and are thus sometimes suitable as a synthetic substitute for common plastics[330-331] as well as having important applications in their own right as biocompatible materials in several biomedical and pharmaceutical applications.[332] For the synthesis of aliphatic polyesters two main strategies are employed, (i) condensation/step growth polymerization or (ii) ring-opening polymerization (ROP). The latter method – as a controlled polymerization method – offers control over molecular weight and polydispersity in comparison to the step growth strategy.

Potential initiators for anionic ROP, which is the basis of the many reports, are nucleophilic reagent such as organo-metals, alkoxides, phosphines, amines, thiols, alcohols and water.[295] The initiator utilized for ring-opening polymerization herein is based on *n*-butyl acrylate macromonomer (BAMM). The terminal olefin can serve as a potentially powerful synthetic handle, for example by a transformation into diols. Thus, the macromonomer has been dihydoxylated through a straightforward transformation method yielding a 1,2-disubstituted macroinitiator for ROP. The overall synthetic strategy is depicted in Scheme 9-5.

The initiating efficiency of an OH terminated macromonomer in the ROP of ε-CL was investigated by the use of various catalytic agents. The initiation efficiency has been established by subjecting the block copolymers pBA-*b*-pCL to SEC, SEC/ESI-MS and LCCC analysis both at the critical conditions of poly(ε-caprolactone) (pCL) and poly(*n*-butyl acrylate) (pBA).

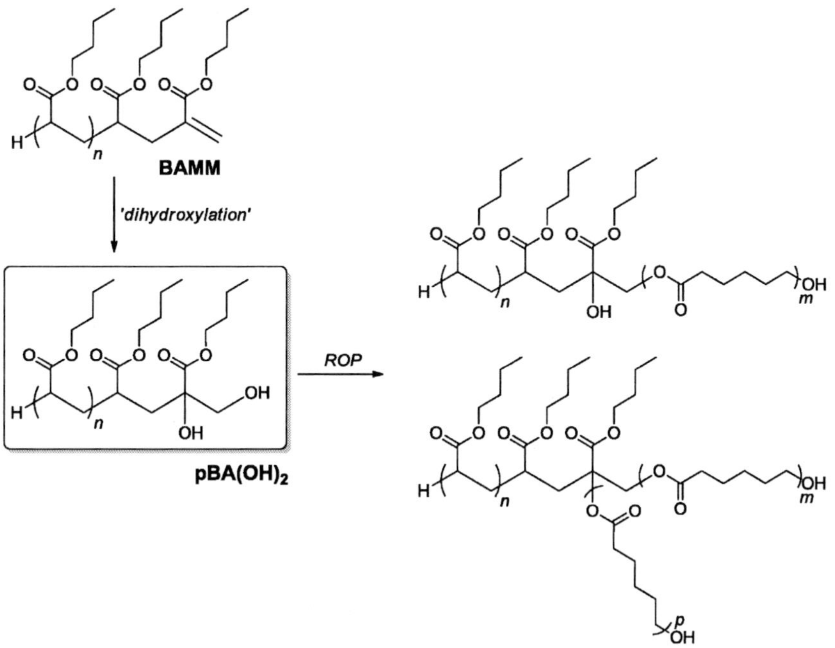

BAMM

'dihydroxylation'

ROP

$pBA(OH)_2$

Scheme 9-5. Synthetic strategy towards block copolymers based on *n*-butyl acrylate macromonomer (BAMM) precursors through the dihydroxylation of BAMM and subsequent ring-opening polymerization of ε-caprolactone. The actual structures of the copolymers differ from the depicted structures as evidenced by SEC/ESI-MS measurements (see Figure 9-12).

9.3.1 Synthesis

*Synthesis of n-Butyl Acrylate Macromonomer **BAMM***

For the macromonomer synthesis *n*-butyl acrylate was polymerized with 5×10^{-3} mol·L^{-1} AIBN in a 5 wt% solution of butyl acetate (285 g, 1.98 mol) (freed from oxygen by purging with argon for approximately 40 min prior to the reaction) in a 2-neck flask at 140 °C. The monomer (15 g, 0.12 mol, 5 wt%) was degassed by purging with argon for close to 20 min in a separate vial sealed airtight with a septum at ambient temperature. The initiator AIBN (5×10^{-3} mol·L^{-1}) was dissolved in butyl acetate, freed from oxygen by purging with argon and added to the reaction mixture at 140 °C. After 5 min the degassed monomer was added to the solvent. The reaction solution was then typically stirred for close to 10 h. After the reaction, the solvent was removed in a vacuum oven at 45 °C. The purity of the synthesized macromonomer was determined by Size Exclusion Chromatography and Electrospray Ionization Mass Spectrometry. **BAMM** – M_n = 1900 g·mol^{-1}, PDI = 1.96.

*Synthesis of the Macroinitiator **pBA(OH)₂***

For the dihydroxylation BAMM (0.35 g, 0.18 mmol, 1 eq) was dissolved in 45 mL ethanol and rigorously stirred at –10 °C. $KMnO_4$ (0.077 g, 0.49 mmol, 2.8 eq) and $MgSO_4$ (0.057 g, 0.47 mmol, 2.7 eq) were dissolved in 45 mL H_2O and added dropwise to the reaction mixture over a period of 1 h. The resulting mixture was stirred for additional 3 h at –10 °C and subsequently allowed to reach ambient temperature. The reaction mixture was filtered off and the residual solvent was extracted with brine (2 × 50 mL) and ethyl acetate (3 × 50 mL). The combined organic phases were dried over $MgSO_4$ and concentrated *in vacuo*. The product was isolated as a yellowish high viscous liquid. The success of the transformation was evidenced by ESI-MS (see Figure 9-3). The macroinitiator pBA$(OH)_2$ was dried *in vacuo* over phosphorous pentoxide prior to the ROP.

*Synthesis of **pBA-b-pCL** via Ring-Opening Polymerization*

The polymerization mixture was prepared inside a glovebox to rigorously exclude water. The initiator pBA$(OH)_2$, catalyst, toluene and ε-caprolactone were added to a pre-dried Schlenk tube, sealed tight and transferred to the outside. The polymerization was carried out at ambient temperature (TBD), 70 °C (Novozym® 435) and 100 °C (Sn$(Oct)_2$), respectively. The polymerization was stopped by cooling the mixture with liquid nitrogen. The crude polymer was precipitated in an excess of cold methanol and dried *in vacuo*. The exact quantities and specifics of the polymerization can be found in Table 9-1.

Table 9-1. Reaction conditions for the ring-opening polymerizations initiated by pBA$(OH)_2$. As catalysts 1,5,7-triazabicyclo[4.4.0]dec-5-ene (TBD, organo-catalysis), Sn$(Oct)_2$ (metal catalysis) and Novozym® 435 (enzyme catalysis) were employed. For each reaction 1.5 mL toluene were used.

catalyst	***n*(ε-CL) /mmol**	***n*(cat) /µmol**	***n*(pBA(OH)₂) /mmol**	***t* /h**	**M_n /g·mol^{-1}**	***PDI***
TBD	0.946	3.6	0.016	2	7300	1.35
Sn(Oct)₂	1.249	3.5	0.030	18	7200	1.43
Novozym® 435	0.946	2 mg	0.008	1.5	5800	1.31

Synthesis of ε-Caprolactone Standards ***pCL***

Polymer standards for the LCCC measurements were synthesized via the ring-opening polymerization method. The polymerization was performed inside a glove box to rigorously exclude water. ε-CL was added to a mixture of 1-butanol and TBD in toluene and stirred for 2 h at ambient temperature and subsequently quenched with benzoic acid. The crude polymer was precipitated in an excess of cold ether : *n*-hexane (1:1). The polymers were analyzed via SEC and SEC/ESI-MS. The molecular weight specifics for the three pCL standards are as follows: **pCL1** – M_n = 6900 g·mol^{-1}, *PDI* = 1.07; **pCL2** – M_n = 13700 g·mol^{-1}, *PDI* = 1.15; **pCL3** – M_n = 50300 g·mol^{-1}, *PDI* = 1.61.

Synthesis of n-Butyl Acrylate Standards ***pBA***

Polymer standards for the LCCC measurements were synthesized *via* the RAFT technique. The initial ratio of $[BA]_0$: $[DoPAT]_0$: $[AIBN]_0$ was 310 : 1 : 0.2. The reaction was performed at 60 °C for variable time intervals. The polymerization was stopped by cooling the mixture with liquid nitrogen. The residual monomer was subsequently removed in a vacuum oven at 45 °C. The polymers were analyzed via SEC and SEC/ESI-MS. The molecular weight specifics for the three pBA standards are as follows: **pBA1** – M_n = 8100 g·mol^{-1}, *PDI* = 1.09; **pBA2** – M_n = 12100 g·mol^{-1}, *PDI* = 1.06; **pBA3** – M_n = 19000 g·mol^{-1}, *PDI* = 1.04.

9.3.2 Results and Discussion

9.3.2.1 Dihydroxylation of BAMM

The transformation of the terminal olefin of the macromonomer synthesized according to the high-temperature synthesis procedure into a macroinitiator for ROP was realized via dihydroxylation. The dihydroxylation procedure was carried out with $KMnO_4$ as the oxidating agent in a one-step procedure.[333] The purity of the resulting OH-terminated macromonomer has been determined via ESI-MS. A typical spectrum of the dihydroxylated macromonomer $pBA(OH)_2$ is depicted in Figure 9-3. The experimental and theoretical mass-to-charge values are listed in Table 9-2 including the accuracy of the assignments indicated by the Δ *m/z* value. The top spectrum of Figure 9-3 depicts two repeat units of the $pBA(OH)_2$ whereas the bottom spectrum represents the starting material BAMM. The chemical structures related to the assignments can be found in Figure 9-2.

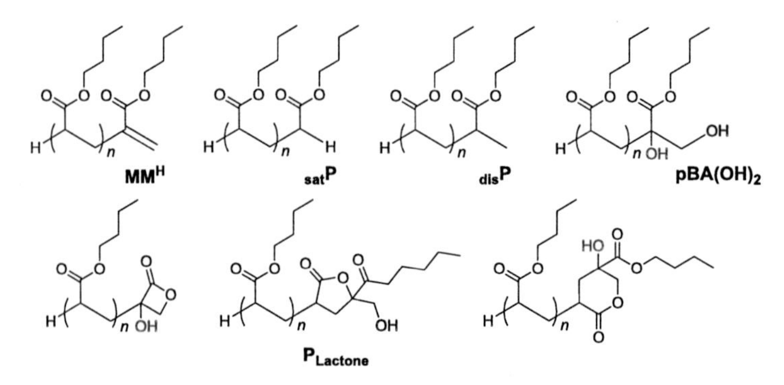

Figure 9-2. Chemical structures of the species appearing in the ESI-MS spectrum of the dihydroxylated product. The lactonization of the diol results in several possible structures depending on the reacting OH functionality (visualized via colored OH).

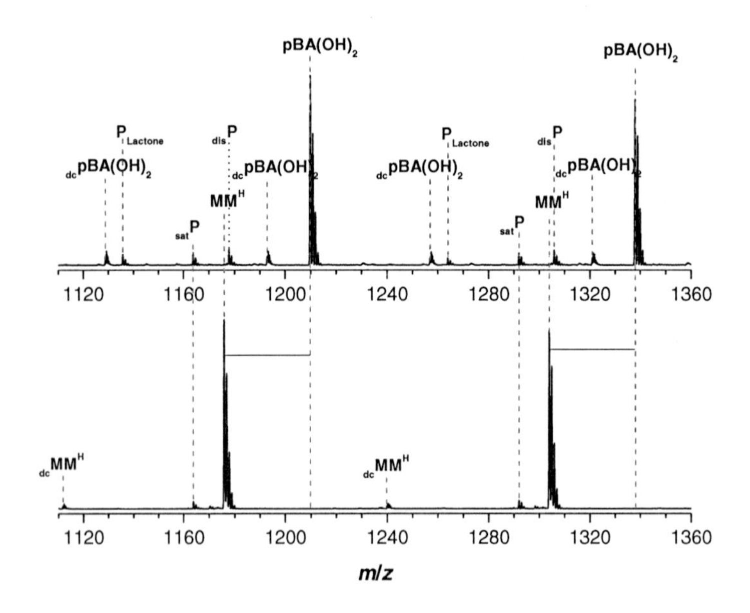

Figure 9-3. ESI-MS spectrum of $pBA(OH)_2$. Two repeat units of the BAMM (bottom) are depicted in comparison to the dihydroxylated product (top). The associated structures of the species are illustrated in Figure 9-2 (dis = product originates from disproportionation, dc = double charged). Reaction conditions: BAMM, $KMnO_4$, $MgSO_4$, ethanol, - 10 °C, 4 h.

Table 9-2. Theoretical and experimental *m/z* ratios of the species of the dihydroxylation of BAMM identified via ESI-MS measurements (see Figure 9-3). Masses have been determined in the *m/z* range 1100–1240. The resolution is close to 0.2 amu.

species	$m/z_{\text{exp.}}$	$m/z_{\text{theo.}}$	$\Delta\, m/z$
$[pBA(OH)_2+2Na]^{2+}$	1128.80	1128.70	0.10
$[P_{Lactone}+Na]^{+}$	1135.80	1135.68	0.12
$[_{sat}P+Na]^{+}$	1163.84	1163.74	0.10
$[MM^{H}+Na]^{+}$	1175.92	1175.74	0.18
$[_{dis}P+Na]^{+}$	1177.84	1177.76	0.08
$[pBA(OH)_2+2Na]^{2+}$	1192.84	1192.75	0.09
$[pBA(OH)_2+Na]^{+}$	1209.88	1209.75	0.13

Inspection of Figure 9-3 clearly evidences that the transformation reaction results in $pBA(OH)_2$ as the main product appearing at $m/z = 1209.88$. The former BAMM is shifted 34 Da to higher masses which is equivalent to the addition of two OH functions to the olefin. The main species can be detected as single and double charged (dc) species, respectively. It should be noted that two minor species are additionally detectable. The appearance of the signal **$P_{Lactone}$** at $m/z = 1135.80$ is associated with an intramolecular cyclization process of the endgroup. The chemical structure of **$P_{Lactone}$** cannot be determined with certainty since the lactonization can occur at different esters on the macromonomer backbone. The resulting structures are isobaric and thus indistinguishable in a mass spectrometric analysis (for chemical structures refere to Figure 9-2). It is well known that methacrylates readily undergo intramolecular cyclization to form lactone or thiolactone endgroups.[334-335] The lactonization originating from the tertiary alcohol functionality resulting in a 5-membered ring structure seems to be more favorable. Since the cyclization can readily be controlled during the transformation process, the **$P_{Lactone}$** species has no influence in the following use as ROP macroinitiator. Beside the observed lactonization product, residual saturated species **$_{dis}P$** formed during the macromonomer synthesis can only be discerned after the dihydroxylation procedure, as in the BAMM starting material it is overlapping (of the isotopic pattern) with the **MM^{H}** signals ($m/z = 1175.74$). The amount of **$_{dis}P$**, which is a termination product of the free radical polymerization emerging from disproportionation, is caused by the high macromonomer synthesis temperature and its quantity cannot be reduced. These saturated species cannot undergo the dihydroxylation and thus hold no initiation properties for ROP. The purification can occur after the polymerization with a simple precipitation step. Additionally to the ESI-MS analysis of the $pBA(OH)_2$, the transformation efficiency of the dihydroxylation has been quantified by ^{1}H-NMR spectroscopy. The geminal protons of the

BAMM show significant resonances in the NMR spectrum between 5 and 7 ppm. After the dihydroxylation of the olefinic terminus the resonances associated with the geminal protons are no more visible indicating their quantitative conversion (the ^{1}H-NMR spectrum is shown in Figure 9-4).

Clearly, the species $pBA(OH)_2$ is formed during the synthesis as the major quantity and therefore presents a potentially excellent ROP macroinitiator of high purity. In the following $pBA(OH)_2$, equipped with a primary and tertiary diol, respectively, has been used as a macroinitiator for the ROP of ε-caprolactone.

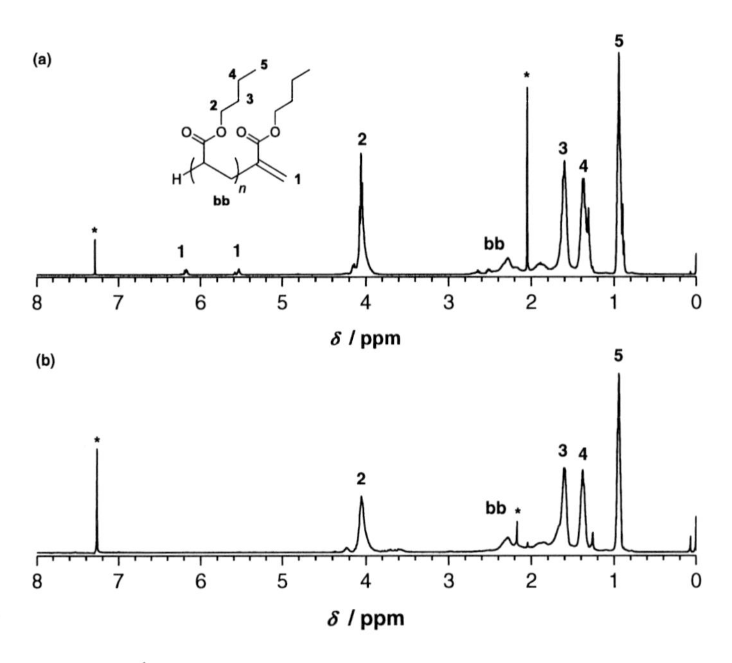

Figure 9-4. 400 MHz ^{1}H-NMR spectrum (in $CDCl_3$) of *n*-butyl acrylate macromonomer (a) and the corresponding dihydroxylated product $pBA(OH)_2$ (b). The disappearance of signal 1 in spectrum (b) evidences full transformation of the vinyl terminus to the diol.

9.3.2.2 $pBA(OH)_2$ Initiated Ring-Opening Polymerization of ε-Caprolactone.

The initiating efficiency of the $pBA(OH)_2$ macroinitiator in ROP has been investigated by changing a range of reaction conditions. Among these, the catalysis system is of critical importance and several are known for ROP, i.e., organo-catalysis, metal catalysis or enzyme mediated ROP. In the present study, we utilized TBD as an organic catalyst, $Sn(Oct)_2$ as the most studied and used metal catalyst and Novozym® 435 as a representative for enzymes employed in ROP.

The polymerization of CL has been carried out in solution of toluene due to viscosity issues of the pure macroinitiator $pBA(OH)_2$. Initially, the organo catalyst TBD was selected for catalysis since this superbase is widely used for ROP.[305,315,336] It is highly active towards any kind of cyclic monomers and easy to remove via precipitation. The initiation efficiency of the macroinitiator $pBA(OH)_2$ was followed by SEC as a basic characterization method. A clear shift from the initial trace related to the macroinitiator to higher molecular weights can be seen in the polymer sample after chain extension with ROP (refer to Figure 9-5).

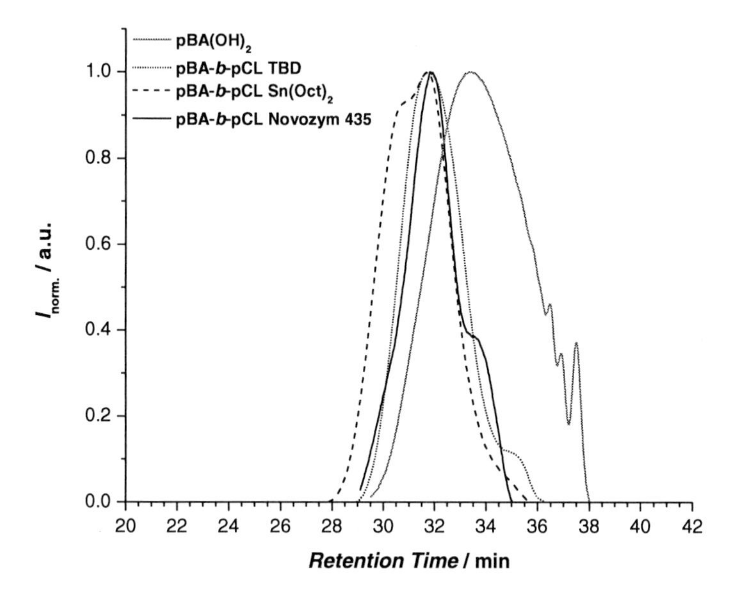

Figure 9-5. SEC chromatograms of the macroinitiator $pBA(OH)_2$ and copolymers pBA-*b*-pCL synthesized via (a) organo-catalysis (1,5,7-triazabicyclo[4.4.0]dec-5-ene, TBD), (b) metal-catalysis ($Sn(Oct)_2$) and (c) enzyme-catalysis (Novozym® 435).

In addition, a slight tailing is visible in the area of higher retention times. The same applies to the polymer synthesized via enzyme catalysis, whereas the tailing at lower molecular weights is higher compared to the organo-catalysis sample. Enzyme catalysts for ROP are quite often employed in the literature and represent a versatile system for the preparation of green and sustainable polymers due to their lower or not present toxicity and renewability.[309,314,316,319,321,337-339] The *in vitro* enzyme catalysis for polymer synthesis has been developed using a lipase enzyme catalyst. In 1993 two independent groups first published the enzyme mediated ROP of CL.[311,313] The SEC chromatogram obtained from the third catalysis system, $Sn(Oct)_2$, shows the same shift to higher molecular weights, however a bimodal molecular weight distribution is obtained. While SEC characterization provides a first indication regarding polymer formation, it entails insufficient information about the chemical structure of the polymers since it cannot be unambiguously established if a block copolymer has been formed.

For a more detailed characterization of the polymer products SEC/ESI-MS spectra have been recorded for the specific samples. The mass spectra of the block copolymer obtained from ROP applying different catalytic systems are depicted in Figure 9-6. For a detailed comparison of the SEC/ESI-MS data, spectra at the retention time *rt* = 14.37 min referring to the triple charged species $[M+3Na]^{3+}$ have been representatively selected.

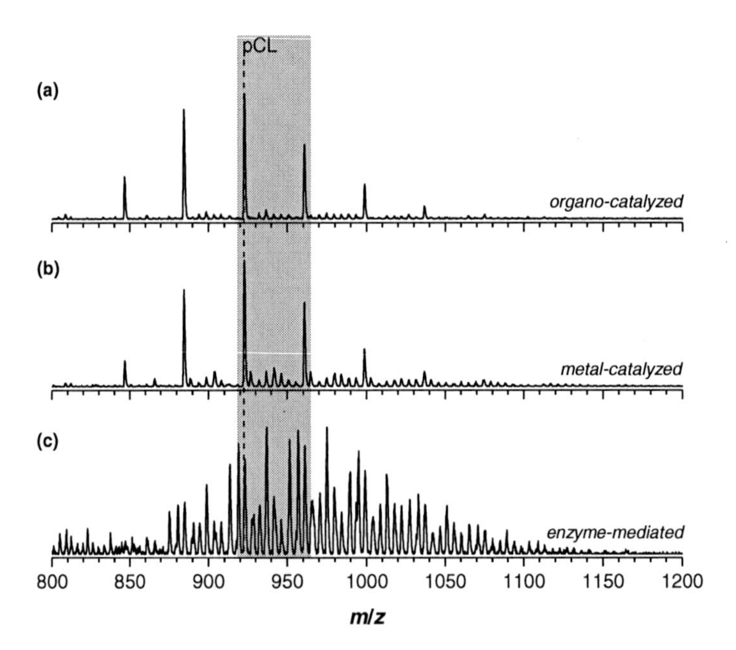

Figure 9-6. Block copolymer formation efficiency of several catalytic reagents investigated via SEC/ESI-MS: (a) organo-catalysis (TBD), (b) metal catalysis ($Sn(Oct)_2$) and (c) enzyme catalysis (Novozym® 435). The gray area displays one repeat unit of pCL. The mass spectra are taken at *rt* = 14.37 min.

Spectrum (a) represents the polymeric material formed under organo-catalysis. The main distribution shows the formation of pCL homopolymer with *n*-butanol endgroups. Several other signals within one repeat unit of the pCL (displayed by the gray area in Figure 9-6) as the main species are recorded, although in minimal quantity. An almost identical spectrum is recorded for the metal catalyzed ROP depicted in spectrum (b). The main product is identical to the previous synthesis, whereas the quantity of block copolymer signals within one repeat unit of pCL is slightly higher. The occurrence of *n*-butanol initiated homopolymer stems from a transesterification reaction, which is a well known side reaction during ROP. Scheme 9-6 illustrates the chemical structures obtained from transesterification side reactions. Once a transesterification step occurs *n*-butanol is released from the macromonomer ester side chain and thus acts as an initiator for ROP.

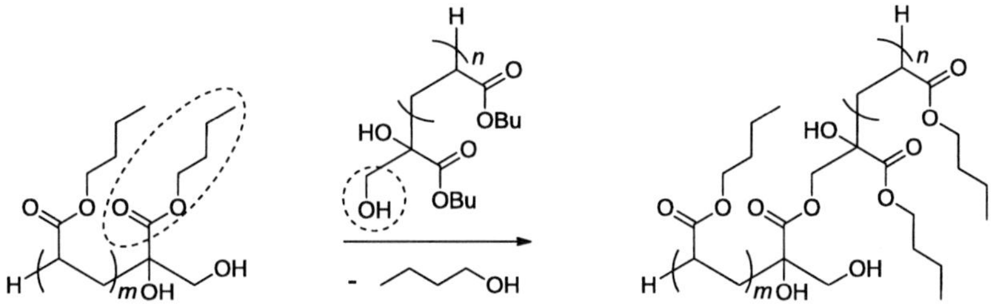

Scheme 9-6. Transesterification during ring-opening polymerization exemplified on one occurring transesterification step. This side reaction can occur randomly on every ester side chain of the macromonomer backbone and from each transesterification *n*-butanol is released as potential ROP initiator.

Besides ROP, both systems – TBD and $Sn(Oct)_2$ – catalyze the inter- or intramolecularly occurring transfer reactions.[340] In the literature, several research groups reported the undesired side reaction during polymerization with organo-catalysis and metal catalysis, which leads to broad polydispersity indices.[302-304,306,336,341-342]

The macroinitiator $pBA(OH)_2$ with a molecular weight of $M_n = 1900$ g·mol^{-1} features (on average) 14 BA repeat units with 14 cleavable *n*-butyl ester side chains, which could potentially undergo the undesired transesterification side reaction during the polymerization. Even at a low catalyst content and reaction time this effect is clearly visible from the results of the SEC/ESI-MS measurements of the 'copolymer'. The formation of pCL homopolymer with *n*-butanol endgroup is significantly higher compared to the formation of desired block copolymer pBA-*b*-pCL. Thus, the above synthetic strategies employing TBD and $Sn(Oct)_2$ as catalyst are not applicable for the chain extension of $pBA(OH)_2$. The polymerization experiments show that the use of organo-catalysis and metal catalysis in ROP initiated by $pBA(OH)_2$ based on *n*-butyl acrylate is unadvisable due to high quantity of transesterification side reactions. However, it should be noted, that the ionization process in mass spectrometric analysis may include discrimination of specific polymer chains. This effect implies that the pCL homopolymers may potentially ionize more efficiently than the copolymeric species resulting in an underestimated initiation efficiency of the $pBA(OH)_2$ species.

While TBD and $Sn(Oct)_2$ did not lead to the desired synthetic outcome, the most efficient block copolymer synthesis was achieved by the use of *in vitro* enzyme catalysis (eROP). The SEC/ESI-MS spectrum (c) (Figure 9-6) gives higher intensity signals referring to the block copolymer in the triple charged state. Compared to spectrum (a) and (b) the efficiency of eROP seems to be higher in forming the block copolymer pBA-*b*-pCL. These initial experiments results were so encouraging that further polymerizations of CL initiated by the macroinitiator $pBA(OH)_2$ have been catalyzed via lipase in an eROP.

Figure 9-7 depicts the mass spectrum of pBA-*b*-pCL at the retention time rt = 14.19-14.43 min synthesized under equivalent eROP conditions as applied for pBA-*b*-pCL of Figure 9-6. For assignment purpose one repeat unit in the region of four-times charged species has been taken (m/z = 780-812). The theoretical and experimental m/z values as well as the accuracy associated with the block copolymer species can be found in Table 9-3.

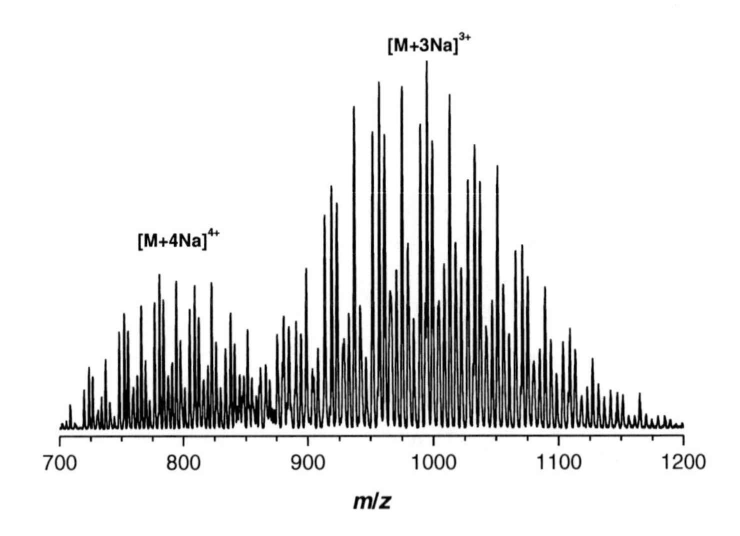

Figure 9-7. SEC/ESI-MS spectrum of pBA-*b*-pCL copolymer synthesized via enzyme catalysis (Novozym® 435). The mass spectrum is taken at rt = 14.19-14.43 min. The exact m/z values associated with each species found in one repeat unit (780-812 m/z) are collated in Table 9-3.

Table 9-3. Exact theoretical and experimental *m/z* ratios associated with the block copolymer pBA-*b*-pCL identified during the SEC/ESI-MS analysis. The label *a-b* represents *a* BA and *b* CL repeat units, respectively. Isobaric structures are collated without repeating the experimental *m/z*. The different indents visualize isobaric structures with different repeat units of BA and CL. Lines with the same indent are differentiated by *a+1* BA units and *b-1* pCL units *etc.*

species BA-CL $[M+4Na]^{4+}$	$m/z_{exp.}$	$m/z_{theo.}$	$\Delta m/z$
pCL 0-26	783.8	783.5	0.3
19-5		783.0	0.8
3-23		783.9	0.1
11-14		783.6	0.2
20-4	787.8	786.5	1.3
4-22		787.5	0.3
12-13		787.0	0.8
21-3	791.0	790.0	1.0
5-21		790.0	1.0
13-12		790.5	0.5
22-2	794.3	793.5	0.8
6-20		794.5	0.1
14-11		793.5	0.8
23-1	797.7	797.0	0.7
7-19		798.0	0.3
15-10		797.5	0.2
8-18	801.3	801.5	0.2
16-9		801.0	0.7
9-17	805.2	805.0	0.2
17-8		804.5	0.7
1-26		805.5	0.3
10-16	809.3	808.5	0.8
18-7		808.0	1.3
2-25		809.0	0.3
pCL 27	812.4	812.0	0.4

species BA-CL $[M+4Na]^{4+}$	m/z exp.	m/z theo.	$\Delta\ m/z$
11-15		812.0	0.4
19-6		811.5	0.9
3-24		812.5	0.1

The interpretation of SEC/ESI-MS spectra of block copolymers is challenging due to the presence of variable block length compositions. It needs to be noted that – even with optimized SEC/ESI-MS settings during the measurement – the isotopic pattern of the species is of medium-resolution (this effect is even more substantial at higher charge states) which leads to a reduced accuracy of the assignments ($\Delta\ m/z$) leading to error margins of up to 1 Da. Molar masses have been used for the calculation of the theoretical m/z values to minimize the broadening effect. For a better understanding of the initiation efficiency one specific repeat unit of BA (128.2 Da) has been analyzed in detail. Inspection of the repeat unit sequence $m/z = 780\text{-}812$ indicates the formation of several block copolymer species. The space between each signal refers to $\Delta\ m/z = [M(\text{BA})\text{-}M(\text{CL})]/z$ (3.5 for $[M+4Na]^{4+}$) according to the difference between 1 BA and +1 CL repeat unit. For instance, the signal at $m/z = 783.8$ represents a block copolymer with a composition of 19 repeat units of BA and 5 repeat units of CL (19-5). The following signal at $m/z = 787.8$ reflects the composition 20-4. The assignments employed here for the determination of the block length are based on the assumption that only the primary alcohol initiates the polymerization. For substantiating this claim organo-catalyzed ROP has been carried out with *tert*-butanol as initiator under identical conditions. Initiation by the tertiary alcohol was observed, however yielding only 15 % pCL whereas a primary homologue alcohol yields 79 % conversion in the same time span. Based on the above conversion difference it seems to be reasonable to expect only minor initiation efficiency of the tertiary alcohol of the macroinitiator $pBA(OH)_2$. The above mentioned m/z = 783.8 can additionally be associated with pCL initiated by *n*-butanol formed from a transesterification side reaction with a repeat unit of 26 as well as the block copolymer species 3-23 and 11-14. Considering a macroinitiator with a specific amount of BA repeat units, i.e., 19-5, the addition of +1 CL unit, i.e., 19-6, is reflected in the signal at $m/z = 812.4$. As Table 9-3 illustrates, more than one species matches with the experimental m/z.

Associated with the *n*-butanol initiated pCL species in the mass spectrum in Figure 9-7 several other species resulting from transesterification can be found in the SEC/ESI-MS spectrum. As mentioned above, transesterification can occur on every random ester side chain of the macromonomer backbone and thus releases the *n*-butanol of the ester moiety (for the chemical structures refer to Scheme 9-6). Table 9-4 summarizes the species associated with transesterification during the polymerization process. The repeat unit picked for the assignments is the same as in Table 9-3. The transesterification is not limited to specific repetition steps which further complicate the assignments of the observed signals. For an in-depth analysis to this side reaction repetitions of the transesterification, i.e., 1, 5 and 10 times on one macromonomer backbone, have been illustrated in Table 9-4.

As already observed for the data provided in Table 9-3, the accuracy of the assignments reflects an error margin of up to 1 Da. However, a few specific structures are more probable than others based on their low $\Delta\, m/z$ value ($\Delta\, m/z \leq 0.3$). For one transesterification step (refer to the columns on the left hand side of Table 9-4) the theoretical and experimental m/z values show a better accuracy compared to the multiple repetition (5 and 10 times as examples shown). The species detected at $m/z = 797.7$, for example, can be assigned to either the copolymer 18-7 ($\Delta\, m/z = 0.3$) with one exchanged ester moiety equal to one graft, 23-4 equal to 5 grafts or 25-2 equal to 10 grafts. The latter two structures are, however, more unlikely due to a $\Delta\, m/z$ of 0.9. Based on these observations, it is probable that transesterifications are limited to a few repetitions. Compared to the assignments for block copolymers without the ester exchange listed in Table 9-3 the assignments including transesterifications (Table 9-4) represent a more accurate structure of the synthesized block copolymers pBA-*b*-pCL. On a structural level, transesterification is resulting in macromonomeric grafts on the actual macroinitiator which leads to a block copolymer with both a linear pCL chain and a grafted pBA chain.

Table 9-4. Exact theoretical and experimental *m/z* ratios associated with transesterifications occurring during polymerization found via SEC/ESI-MS analysis. For illustration purpose 1, 5 and 10 times transesterification are illustrated. The label *a-b* represents *a* BA and *b* CL repeat units, respectively. Isobaric structures are collated without repeating the experimental *m/z*. The species are found in charged state $[M+4Na]^{4+}$.

	1 × transesterification			5 × transesterification			10 × transesterification		
m/z exp.	**species BA-CL**	***m/z*** theo.	**Δ *m/z***	**species BA-CL**	***m/z*** theo.	**Δ *m/z***	**species BA-CL**	***m/z*** theo.	**Δ *m/z***
783.8	14-11	784.0	0.1	19-8	784.5	0.7	21-9	784.5	0.7
	22-2	783.5	0.4						
	6-20	784.4	0.6						
787.5	7-19	788.0	0.5	20-7	788.0	0.5	22-8	788.0	0.5
	15-10	787.5	0.0	12-16	788.5	1.0			
791.0	16-9	791.0	0.0	13-15	792.0	1.0	23-7	791.5	0.5
	8-18	791.5	0.5	21-6	791.5	0.5			
794.3	9-17	795.0	0.6	22-5	795.0	0.7	24-6	795.0	0.7
	17-8	794.5	0.1						
797.7	18-7	798.0	0.3	23-4	798.5	0.9	25-2	798.6	0.9
	10-16	798.5	0.8						
801.3	19-6	801.5	0.3	24-3	802.0	0.8	26-4	802.1	0.8
	11-5	802.0	0.7						
805.2	20-5	805.0	0.2	25-2	805.5	0.4	27-3	805.6	0.4
	12-14	805.5	0.3	17-11	806.0	0.9			
809.3	13-13	809.0	0.3	26-1	809.0	0.2	20-11	809.6	0.3
	21-4	808.5	0.7	18-10	809.5	0.3	28-2	809.1	0.2
				10-19	810.0	0.8			
812.4	14-12	812.5	0.1	19-9	813.0	0.6	29-1	812.6	0.2
	22-3	812.0	0.4	11-18	813.5	1.1	21-10	813.1	0.6

SEC/ESI-MS represents a versatile characterization method for the determination of block copolymers and constitutes the basis for understanding the initiation efficiency. As discussed above, pCL homopolymer from transesterification is still present and indistinguishable from the isobaric copolymer species. It is well known from the literature that eROP can also lead to transesterification side reactions – albeit to a much lesser extent than other catalytic systems – if cleavable ester bonds are present.[310,314,317-318,339,343-344] Thus, further characterization methods have been employed in the present study to obtain a detailed understanding of the eROP efficiency of the $pBA(OH)_2$ macroinitiator. For the separation of homo- and block copolymers the synthetic products have been subjected to LCCC analysis.

9.3.2.3 Characterization of pBA-*b*-pCL via LCCC

In the current study the critical conditions of pCL *and* pBA were applied to shed further light on the initiation efficiency of the macroninitiator $pBA(OH)_2$ in eROP. The critical conditions of pCL have been realized on two reversed phase columns in series with THF/MeOH 20/80 (v/v) as the mobile phase. The pCL standards ranging from 6900 to 50100 $g{\cdot}mol^{-1}$ elute at an elution volume of V_e = 4.92 mL. The pBA standards however are adsorbed on the stationary phase at these conditions and elute in LAC mode, i.e., the elution volume of the pBA samples of different block length increases with increasing molecular weight. Figure 9-8 displays the chromatographic characterization of eROP synthesized pBA-*b*-pCL at the critical conditions of pCL. The block copolymer (straight line), BAMM (dotted line) and the dihydroxylated macroinitiator $pBA(OH)_2$ (dashed line) have been measured to facilitate the assignments. The black line results from the injection of the pBA-*b*-pCL, yielding a multimodal elugram. Inspection of Figure 9-8 suggests the formation of several species during the polymerization process. The first peak (1) detected at an elution volume of V_e = 4.92 mL is appearing at the critical point of the pCL standards and matches with pCL homopolymer formed due to transesterification reactions. The following distribution (2) at an elution time of V_e = 5.09 mL is associated with unreacted macroinitiator $pBA(OH)_2$, since the pure macroinitiator elutes at the same elution volume. Even with two columns in series it is not possible to clearly separate the macroinitiator $pBA(OH)_2$ from the pCL homopolymers. Nevertheless, two species are distinguishable and relate to the reference measurements of the pCL homopolymers and residual $pBA(OH)_2$ macroinitiator (dashed line Figure 9-8). The additional peak (3) of the macroinitiator $pBA(OH)_2$ distribution represents residual species $_{\mathbf{dis}}\mathbf{P}$ which is already formed during the high-temperature synthesis of the BAMM and becomes visible after the dihydroxylation procedure. Due to the purification of the crude copolymer, the species $_{\mathbf{dis}}\mathbf{P}$ is separated and thus detectable during the LCCC measurement of pBA-*b*-pCL. The true copolymer pBA-*b*-pCL is represented by the following broad signal (4) starting at V_e = 6.26 mL.

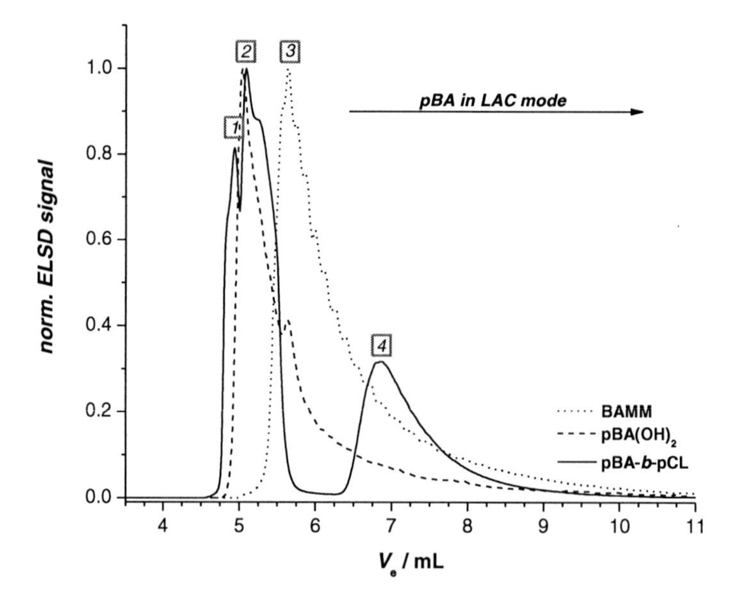

Figure 9-8. LCCC elugrams of starting material BAMM (dotted line), macroinitiator $pBA(OH)_2$ (dashed line) and the block copolymer pBA-*b*-pCL (black line) at the critical conditions of pCL on a series of two reverse phase columns (the dashed straight line at V_e = 4.92 mL represents the elution volume of the pCL standards). Parts of the elugram have been labeled (1-4) for a better understanding (for details see text).

The presence of a broad signal confirms the existence of BA in the species since its adsorption on the stationary phase is clearly visible. The knowledge obtained from the LCCC measurements at the critical conditions of pCL confirms the first results from SEC/ESI-MS measurements. Besides the eROP process, lipase catalyzes the transesterification reaction, yielding pCL homopolymer with *n*-butanol endgroups, which was postulated on the basis of the SEC/ESI-MS measurements due to the appearance of the isobaric species related to pCL homopolymer (the pCL standards applied for determination of the critical conditions are identical to the *n*-butanol initiated pCL species from transesterification). In addition, the observation of residual macroinitiator $pBA(OH)_2$ leads to the conclusion that the initiation efficiency is low compared to the case when small initiators with lower demand on the sterical periphery are employed. Longer reaction times could improve the efficiency yet also increase the probability of transesterification side reactions. Consequently, the amount of residual macroinitiator $pBA(OH)_2$ would be decreased and, simultaneously, the amount of pCL homopolymer is increased. Indeed, polymerizations with longer reaction times have been carried out resulting in less quantity of block copolymer with the same amount of residual macroinitiator.

Apart from eROP copolymer samples, the efficiency of the organo-catalysis and metal catalysis systems has been subjected to SEC/ESI-MS. To confirm the first results, the copolymers synthesized via the above mentioned catalysis systems have been underpinned by LCCC at the critical conditions of pCL (for the elugrams refer to Figure 9-9).

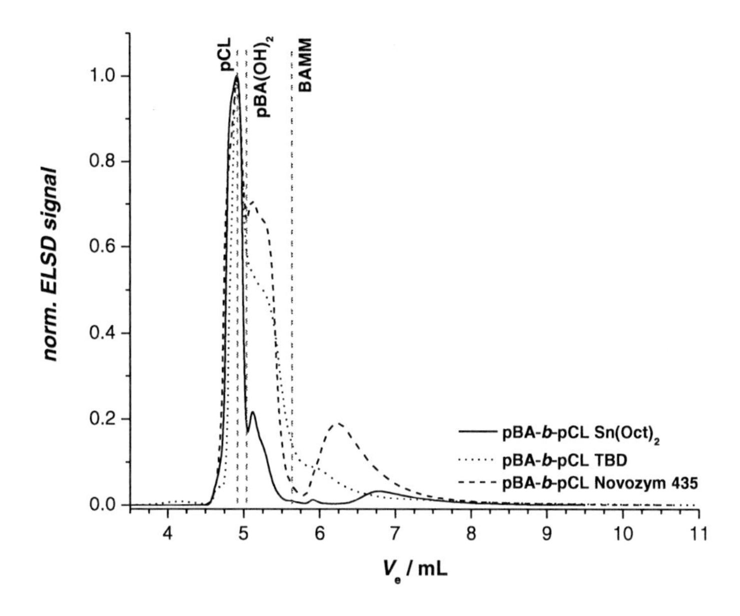

Figure 9-9. LCCC elugrams of pBA-*b*-pCL catalyzed via 1,5,7-triazabicyclo[4.4.0]dec-5-ene (TBD, dotted line), $Sn(Oct)_2$ (straight line) and Novozym® 435 (dashed line) at the critical conditions of pCL on a series of two reverse phase columns (the dashed straight gray lines represent the elution volume of the pCL standards, $pBA(OH)_2$ and BAMM, respectively).

Similar to the eROP copolymer measurements, each sample shows residual macroinitiator $pBA(OH)_2$ as well as homo- and copolymeric species. The observation of minor amounts of macroinitiator – expressed by a less intense signal compared to the homopolymer signal – illustrates a higher initiation efficiency compared to the eROP system. However, a significant amount of pCL homopolymers is detected, as transesterification side reactions are highly favored during organo-catalysis and metal catalysis. A broadened signal at higher elution volume reflects minor amounts of the desired copolymer pBA-*b*-pCL. LCCC measurements confirm the earlier observations of the SEC/ESI-MS measurements: block copolymer formation is most efficient with eROP by lipase catalysis.

The block copolymer sample synthesized via eROP already measured at the critical conditions of pCL (refer to Figure 9-8) has additionally been analyzed via LCCC at the critical conditions of pBA. The stationary phase has been kept unchanged (i.e., a reverse phase column) yet reduced to one column due to strong adsorption effects of pBA on the series of reversed phase columns employed at the critical conditions of pCL. Critical conditions were found to be at a mobile phase composition of THF/MeOH 40/60 (v/v). Here, the pCL standards elute in SEC mode, i.e., decrease in elution volume with increase of molecular weight. Figure 9-10 represents the chromatographic results of the block copolymer pBA-*b*-pCL obtained at the critical conditions of pBA.

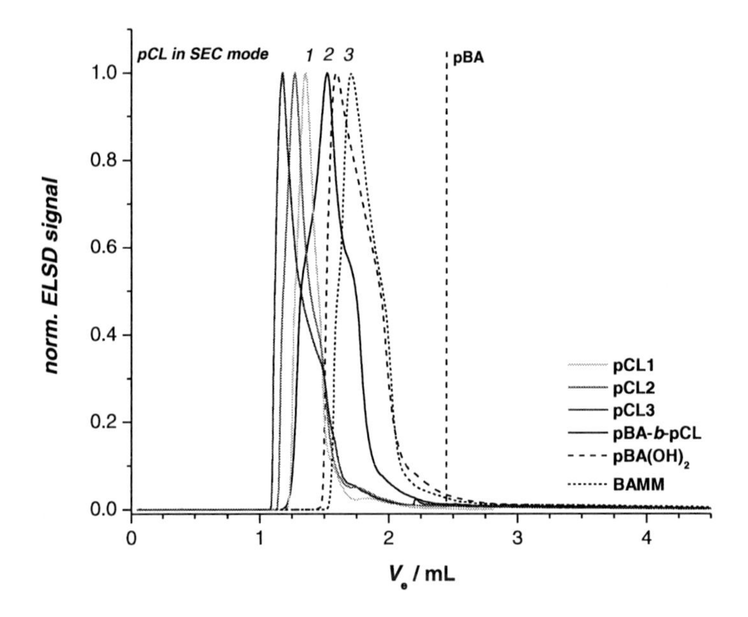

Figure 9-10. LCCC elugrams of pCL standards (gray scale) in SEC mode, starting material BAMM (dotted black line), initiator $pBA(OH)_2$ (dashed black line) and the block copolymer pBA-*b*-pCL (black line) at the critical conditions of pBA on a reverse phase column (the dashed straight line at $V_e = 2.44$ mL represents the elution volume of the pBA standards). Parts of the elugram have been labeled (1-3) for a better understanding (for details see text).

The pCL standards are depicted in gray scale eluting at $V_e = 1.18$ mL (pCL3), $V_e = 1.27$ mL (pCL2) and $V_e = 1.36$ mL (pCL1). It is clearly visible that the elution occurs in SEC mode. Again, both macromonomer BAMM and macroinitiator $pBA(OH)_2$ have been measured to verify the assignments of the elugrams. The elugram of the copolymer sample (black line) appears to show a multimodal distribution as was observed vice versa at the critical conditions of pCL. By comparison with the elugrams of the macroinitiator $pBA(OH)_2$ and the pCL standards the multimodality can be understood: The first shoulder (1) of the elugram at approximately

V_e = 1.35 mL can be identified as pCL homopolymer from transesterification side reactions, the second – and most intense – signal (2) of the elugram at V_e = 1.52 mL is related to the actual copolymer pBA-*b*-pCL and is followed by a further shoulder (3) visible at approximately V_e = 1.68 mL related to residual macroinitiator $pBA(OH)_2$. Such an outcome is congruent with the previous measurements at the critical conditions of pCL.

Previously, Chagneux *et al.* described the characterization of pBA-*b*-pCL block copolymers via LCCC at the critical conditions of pBA and pCL, respectively.[117] For the block copolymer synthesis a dual initiator has been employed for ROP of CL and nitroxide-mediated polymerization (NMP) for BA. The ROP has been performed with $Sn(Oct)_2$ catalysis in bulk or toluene in a temperature range between 80 and 120 °C. LCCC measurements at the critical conditions of pCL indicated the formation of pCL species with no residual macroinitiator pBA-OH. Herein, the elugram shifted to lower elution time could be related to either pCL or pBA-*b*-pCL. Further LCCC measurements at the critical conditions of pBA have been conducted to consolidate these observations. The elugram revealed a peak at the elution time of the pBA standards which could be related to either pBA-*b*-pCL or the precursor pBA-OH. Additionally, the formation of small quantities of pCL homopolymer was observed. By comparison of both measurements, the authors conclude that the formation of pBA-*b*-pCL block copolymers is feasible with high initiation efficiency, and pCL homopolymer formation initiated by water is a side product. Surprisingly, transesterification side reactions are not mentioned in their work, although the pCL could potentially be initiated by *n*-butanol released from transesterifications.

For a better assignment of the occurring species during LCCC measurements, the elugrams obtained via evaporative light-scattering detection (ELSD) of the block copolymer sample measured at the critical conditions of pCL and pBA, respectively, (refer to the black elugram of Figure 9-8 and Figure 9-10) have been deconvoluted with an Gaussian function using the PeakFit Program (V4.12, SeaSolve Software Inc.). The deconvolution of the elugrams is depicted in Figure 9-11.

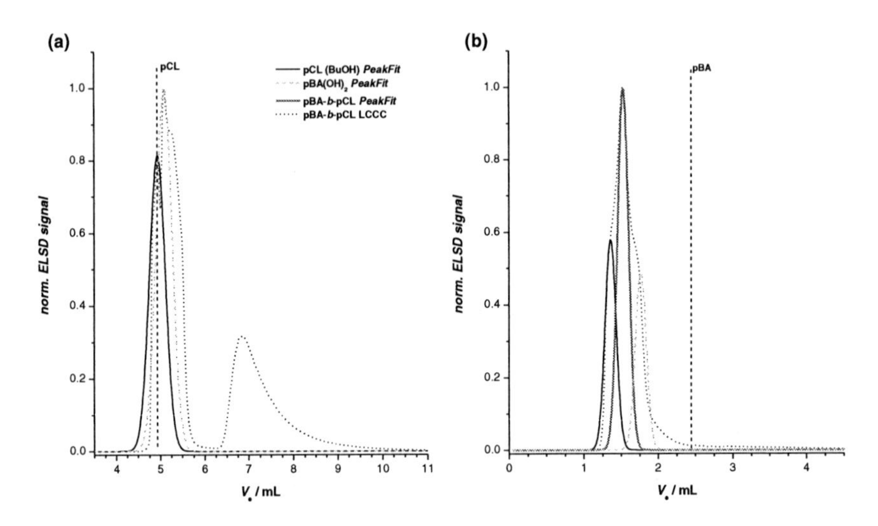

Figure 9-11. LCCC trace deconvolution. The elugram of pBA-*b*-pCL (dotted line) at the critical conditions of (a) pCL and (b) pBA in comparison to the integration of the deconvoluted peaks estimating the quantity of the detected species of pBA-*b*-pCL (bold gray curve), pCL from transfer reactions (black curve) and the residual initiator $pBA(OH)_2$ (dashed gray curve).

The elugrams have been deconvoluted according to their shape representing the residual macroinitiator $pBA(OH)_2$ (dashed gray curve), homopolymer pCL from transesterification (black curve) and block copolymer pBA-*b*-pCL (bold gray curve), respectively. The denconvolution is utilized to defragment the multimodal distributions, which is why the broadened peak related to pBA-*b*-pCL (Figure 9-11 (a)) has been excluded from the deconvolution process. The algorithm was executed iteratively until the correlation coefficient of the fitted data was minimized (via the least squares method) and the results fitted to the raw experimental data.[119] The potential drawback of this method is based on the possible non-linearity of the ELSD signal between intensity and molecular mass for different polymer types and solvents.[345-346] In the current situation different macromolecular species are used, i.e., pBA and pCL with different particle size and thus different droplet size. Nevertheless, deconvolution is used as a versatile tool to visualize the quantity of various species detected at critical conditions of both pCL and pBA. Figure 9-11 shows on the left hand side (a) the elugram at the critical conditions of pCL and at the critical conditions of pBA on the right hand side (b). Deconvolution of the elugrams resulted in isolated peaks referring to the following species: residual macroinitiator $pBA(OH)_2$ (dashed gray curve), homopolymer pCL from transesterification (black curve) and block copolymer pBA-*b*-pCL (bold gray curve and dotted curve from raw data), respectively. However, the exact quantity of block copolymer species and thus macroinitiation efficiency cannot be determined, due to potential non-linearity effects of the ELSD. The deconvolution should be understood as a visualization tool demonstrating the hidden distributions of a multimodal LCCC ELSD signal.

9.3.3 Conclusions

The BAMM macromonomer synthesized via the high-temperature acrylate polymerization has been successfully transformed into a macroinitiator for ring-opening polymerization by dihydroxylation of the olefinic terminus via a facile one-pot transformation yielding hydroxyl functionalized macromonomer $pBA(OH)_2$. The purity of the macroinitiator $pBA(OH)_2$ has been established via ESI-MS characterization. The initiation efficiency of $pBA(OH)_2$ in a ROP of ε-CL has been investigated by applying organo-catalysis (TBD), metal catalysis ($Sn(Oct)_2$) and enzyme catalysis (Novozym® 435). The latter catalysis system emerged as the most efficient of the three, however with transesterification side reaction occurring on the macromonomer backbone. The block copolymer formation processes have been inspected via SEC, SEC/ESI-MS and LCCC at the critical conditions of pCL and pBA, respectively. Data obtained from SEC/ESI-MS measurements indicate a chemical structure deviating from the theoretical expectations. The proposed structure combines linear pCL chains as well as pBA chains with macromonomeric grafts. Both structural parts are linked through the diol of the macroinitiator (see Figure 9-12).

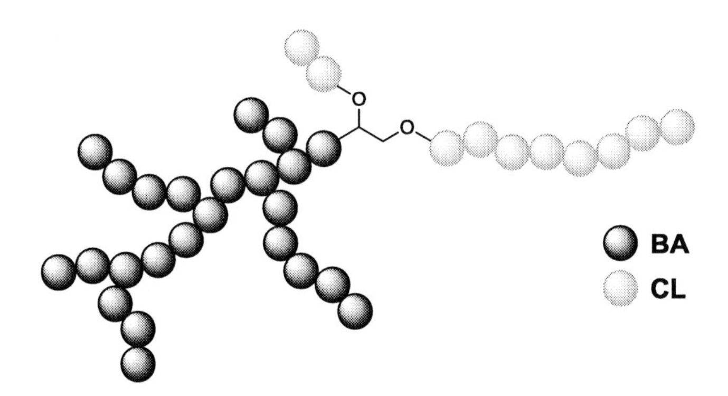

Figure 9-12. Proposed schematic structure of the copolymers pBA-*b*-pCL obtained from eROP initiated by the macroinitiator $pBA(OH)_2$ and evidenced by SEC/ESI-MS and LCCC measurements.

The primary OH functionality acts as the major initiator (polymer chain on the right side of Figure 9-12), whereas the tertiary OH on the macroinitiator slowly adds CL monomer or even remains inactive. The structural difference cannot be distinguished in either SEC/ESI-MS or LCCC due to the isobaric nature of both structures. The access route to block copolymers via eROP of macroinitiator $pBA(OH)_2$ via a two step synthesis based on high-temperature polymerization macromonomer although certainly providing pBA-*b*-pCL, is thus challenged by side reactions such as transesterification of the acrylate backbone.

10

Materials and Characterization

10.1 Materials

10.1.1 Chemicals Used in Chapter 6

Methyl acrylate (MA, Aldrich, 99%), ethyl acrylate (EA, Aldrich, 99%), butyl acrylate (BA, Fluka, 99%), *tert*-butyl acrylate (*t*BA, Aldrich, 99%), 2-[[(butylamino)carbonyl]oxy]ethyl acrylate (BECA, Aldrich, 99%), 2-ethylhexyl acrylate (2-EHA, Aldrich, 98%), isobornyl acrylate (iBoA, Aldrich, technical grade) were deinhibited by percolating over a column of activated basic alumina (Acros, Brockmann I, for chromatography, 50-200 μm). Hexyl acetate (Acros, 99%) was used as received. The initiators 1,1'-azobis(isobutyronitrile) (AIBN, Acros, 98%) and 1,1'-azobis(cyclohexanecarbonitrile) (VAZO 88) were recrystallised twice from ethanol prior to use.

10.1.2 Chemicals Used in Chapter 7

Hexyl acetate (99%) and palladium on carbon (10% Pd, unreduced) were purchased from Acros and used as received. 2,2'-Azobis(isobutyronitrile) (AIBN, Sigma-Aldrich, 98%) was recrystallized twice from methanol and ethyl acrylate (Aldrich, 99%) was deinhibited over basic alumina (Acros, Brockmann I, for chromatography, 50-200 μm) prior to use. 9-Bromo-1-nonanol (95%), 6-bromo-1-hexanol (97%), acryloyl chloride (99%),

4-(dimethylamino)pyridine (DMAP, 99%), hydroquinone (99+%) and $CuSO_4 \cdot 5H_2O$ (98+%) were purchased from Sigma Aldrich and used as received. NaN_3 (Riedel de Haen), propargyl alcohol (99%, Fluka) and (+)sodium L-ascorbate (Sigma) were all used as received. Acetonide-protected 2,2-bis(hydroxymethyl)propionic anhydride (An-bis-MPA) was synthesized according to the literature[261] from 2,2-bis(hydroxymethyl)propionic acid (bis-MPA), kindly donated by Perstop AB, Sweden. DOWEX® 50W-X2 50-100 (H^+) ion-exchange resin (Alfa Aesar) was purified with methanol prior to use. Dimethylsulfoxide (DMSO), triethylamine, dichlormethane (CH_2Cl_2), methanol and the TLC plates (TLC silica gel 60 F_{254}) were purchased from Merck and used as received. Pyridine, tetrahydrofuran (THF) and diethylether were all AnalaR Normapur and received from VWR. Ethyl acetate (EtOAc) and n-heptane were p.a. grade and obtained from Fisher Scientific. Column chromatography was performed with Acros Organics silica gel pore size 60 Å, 40-60µm.

10.1.3 Chemicals Used in Chapter 8

The monomers *n*-butyl acrylate (BA, Fluka, 99%) and benzyl acrylate (BzA, Alfa Aesar, 97%) were deinhibited by percolating over a column of activated basic alumina (Acros, Brockmann I, for chromatography, 50-200 µm) prior to use. Hexyl acetate (Acros, 99%) was used as received. The initiator 1,1'-azobis(isobutyronitrile) (AIBN, Sigma Aldrich, 98%) was recrystallized twice from ethanol prior to use. 2-[(dodecylsulfanyl)carbonothioyl]sulfanyl propanoic acid (DoPAT) is of a purity of >99% evidenced by nuclear magnetic resonance (NMR) and is obtained from Orica Pty Ltd., Melbourne, Australia as a donation and used as received. Toluene (extra dry, water <30ppm, Acros Organics), methanol (VWR, Prolabo, Normapur), tetrahydrofuran (multisolvent, 250ppm BHT, Scharlau) and *n*-hexane (rotisolv, HPLC grade, Roth) were used as received.

10.1.4 Chemicals Used in Chapter 9

The monomer *n*-butyl acrylate (BA, Fluka, 99%) was deinhibited by percolating over a column of activated basic alumina (Acros, Brockmann I, for chromatography, 50-200 µm) prior to use. 2-[(Dodecylsulfanyl)carbonothioyl]sulfanyl propanoic acid (DoPAT) is of a purity of >99% evidenced by nuclear magnetic resonance (NMR) and is obtained from Orica Pty Ltd., Melbourne, Australia as a donation and used as received. The initiator 1,1'-azobis(isobutyronitrile) (AIBN, Sigma Aldrich, 98%) was recrystallized twice from ethanol prior to use. ε-Caprolactone (Alfa Aesar, 99%) was distilled from CaH_2 and kept over molecular sieve. *n*-Butanol (Acros Organics, 99+%, AcroSeal®, extra dry) was used as received. Diethyl ether, ethyl acetate, butyl acetate and ethanol were purchased from VWR (Prolabo, Normapur) and also used as received. The catalysts 1,5,7-triazabicyclo[4.4.0]dec-5-ene (TBD, Sigma Aldrich) and tin(II) 2-ethylhexanoate ($Sn(Oct)_2$, Alfa Aesar, 96%) were used as received. Lipase from *Candida antarctica* within a macroporous acrylic resin

(Novozym® 435) was purchased from Sigma Aldrich and dried *in vacuo* over phosphorous pentoxide prior to use. Potassium permanganate (Acros Organics), magnesium sulfate (Roth), toluene (Acros Organics, 99.85%, extra dry over molecular sieve, water <30ppm), tetrahydrofuran (Scharlau, multisolvent, 250ppm BHT) and methanol (Fisher Scientific, HPLC grade) were used as received.

10.2 Characterization

10.2.1 ^{1}H and ^{13}C Nuclear Magnetic Resonance Spectroscopy (NMR)

The purity of the synthesized compounds was confirmed via ^{1}H- and ^{13}C-NMR spectroscopy. The spectra were recorded on a Bruker AM250 at 250 MHz and on a Bruker AM400 spectrometer at 400 MHz for hydrogen nuclei and 100 MHz for carbon nuclei. Samples were dissolved in acetone-d_6, $CDCl_3$ and MeOD, respectively. The solvent signal was used as an internal standard. In Chapter 8, integration of the CH_2 (3) signal of the macromonomer was compared to the H_{ar} (2) and the CH_2 (1) signal of the benzyl acrylate, respectively, for copolymerization composition determination.

10.2.2 Size Exclusion Chromatography (SEC)

Size Exclusion Chromatography (SEC) measurements were performed on a Polymer Laboratories (Varian) PL-GPC 50 Plus Integrated System, comprising an autosampler, a PLgel 5µm bead-size guard column (50 × 7.5 mm) followed by one PLgel 3µm Mixed E column (300 × 7.5 mm), three PLgel 5µm Mixed C columns (300 × 7.5 mm) and a differential refractive index detector using THF as the eluent at 35 °C with a flow rate of 1 mL·min^{-1}. The SEC system was calibrated using linear poly(styrene) standards ranging from 160 to 6×10^{6} g·mol^{-1} ($K = 14.1 \times 10^{-5}$ dL·g^{-1} and $\alpha = 0.70$).[347] The resulting molecular weight distributions have been universally recalibrated using Mark-Houwink parameters for poly(*n*-butyl acrylate) ($K = 12.2 \times 10^{-5}$ dL·g^{-1}, $\alpha = 0.70$),[290] poly(*tert*-butyl acrylate) ($K = 19.7 \times 10^{-5}$ dL·g^{-1}, $\alpha = 0.66$), poly(isobornyl acrylate) ($K = 5.00 \times 10^{-5}$ dL·g^{-1}, $\alpha = 0.745$),[348] poly(methyl acrylate) ($K = 8.35 \times 10^{-5}$ dL·g^{-1}, $\alpha = 0.768$), poly(ethyl acrylate) ($K = 8.34 \times 10^{-5}$ dL·g^{-1}, $\alpha = 0.767$), poly(2-ethylhexyl acrylate) ($K = 19.6 \times 10^{-5}$ dL·g^{-1}, $\alpha = 0.628$),[349] and poly(ε-caprolactone) ($K = 13.95 \times 10^{-5}$ dL·g^{-1}, $\alpha = 0.786$).[350] For the pBA-*co*-pCL block copolymers, the Mark-Houwink parameters for poly(ε-caprolactone) and for the pBAMM-*co*-pBzA copolymers the Mark-Houwink parameters for poly(*n*-butyl acrylate) have been employed. The concentration of the injected polymer solution was close to 2 mg·mL^{-1}.

10.2.3 Size Exclusion Chromatography Electrospray Ionization – Mass Spectrometry (SEC/ESI-MS)

Mass spectra were recorded on a LXQ mass spectrometer (ThermoFisher Scientific) equipped with an atmospheric pressure ionization source operating in the nebulizer-assisted electrospray mode. The instrument was calibrated in the *m/z* range 195-1822 using a standard containing caffeine, Met-Arg-Phe-Ala acetate (MRFA) and a mixture of fluorinated phosphazenes (Ultramark 1621) (all from Aldrich). A constant spray voltage of 4.5 kV and a dimensionless sweep gas flow rate of 2 (approx. 3 L·min^{-1}) and a dimensionless sheath gas flow rate of 12 (approx. 1 L·min^{-1}) were applied. The capillary voltage, the tube lens offset voltage and the capillary temperature were set to 60 V, 110 V and 275 °C respectively. The solvent was a 3/2 (v/v) mixture of THF : methanol with polymer concentration ≈ 0.2 mg·mL^{-1}. The instrumental resolution of the employed experimental set up is 0.1 amu.

For SEC/ESI-MS the LXQ was coupled to a Series 1200 HPLC system (Agilent) that consisted of a solvent degasser (G1322A), a binary pump (G1312A) and a high performance autosampler (G1367B), followed by a thermostatted column compartment (G1316A). Separation was performed on two mixed bed GPC columns (Polymer Laboratories, Mesopore 250 × 4.6 mm, particle dia. 3µm) with pre-column (Mesopore 50 × 4.6 mm) operating at 30 °C. THF at a flow rate of 0.3 mL·min^{-1} was used as the eluent. The mass spectrometer was coupled to the column in parallel to a RI detector (G1362A with SS420 × A/D) in a setup described previously.[351] A 0.27 mL·min^{-1} aliquot of the eluent was directed through the RI detector and 30 µL·min^{-1} infused into the electrospray source after postcolumn addition of a 0.1 mM solution of sodium iodide in methanol at 20 µL·min^{-1} by a micro flow HPLC syringe pump (Teledyne ISCO, Model 100DM). The polymer solutions (20 µL) with a concentration of 2 mg·mL^{-1} were injected into the HPLC system.

10.2.4 Matrix-Assisted Laser Desorption and Ionization (MALDI) – Time of Flight (ToF) Mass Spectrometry

The MALDI-ToF MS spectra were recorded on a Bruker UltraFlex MALDI-ToF equipped with a SCOUT-MTP Ion Source (Bruker Daltonics, Bremen), a N_2-Laser (337 nm), a gridless ion source and reflector design. All spectra were acquired using a reflector positive method with an acceleration voltage of 25 kV and a reflector voltage of 26.3 kV. As matrix material 2,5-dihydroxybenzoic acid (DHB) was used as a 0.1 M solution in THF and sodium trifluoroacetate (1 mg·mL^{-1}, Aldrich, 98%) was added to each solution. For sample preparation 5 µL of a 5 µM sample solution in THF was added to 20 µL of the matrix solution. 1 µL of this solution was applied to a spot of the target plate and the solution was left to crystallize at ambient temperature. The achieved spectra were analyzed with the FlexAnalysis Bruker Daltonics software (Version 2.2).

10.2.5 Liquid Chromatography at Critical Conditions (LCCC)

The analysis of the copolymers was performed under critical conditions using an Agilent SECcurity SEC system (1200 series), comprising a degasser (G1379B), binary pump (G1312A), autosampler (G1329A) and a temperature controlling unit (G1316A), equipped with the corresponding columns (for the specifics of the critical conditions refer to Table 10-1) at 35 °C. Detection was carried out with an evaporative light scattering detector (ELSD 400, SofTA Corporation). The flow rate of the eluent mixture (specifics below) was set to 0.5 $mL \cdot min^{-1}$ and the samples were prepared in the mentioned eluent at a polymer concentration of 2 $mg \cdot mL^{-1}$, where 25 µL of the sample were injected.

Data acquisition and processing were accomplished with the PSS WinGPC Unity software package from Polymer Standards Service (PSS, Mainz, Germany) and the Agilent ChemStation for LC 3D Systems (Rev. B.04.01) from Agilent Technologies.

Table 10-1. Summary of the critical conditions applied within this work. The critical conditions were found at 35 °C at a flow rate of 0.5 $mL \cdot min^{-1}$.

polymer	stationary phase[a]	mobile phase
poly(*n*-butyl acrylate)	*Nucleosil C18* [b]	THF/*n*-hexane
	300 Å - 5 µm	28.7/71.3 (v/v)
	250/4.6 mm	
	Zorbax Eclipse XDB-C18 [c]	THF/MeOH
	100 Å - 5 µm	40/60 (v/v)
	150/4.6 mm	
poly(ε-caprolactone)	*Zorbax Eclipse XDB-C18* [c]	THF/MeOH
	100 Å - 5 µm	20/80 (v/v)
	150/4.6 mm	
	Pack ODS-A	
	300 Å - 5 µm	
	250/4.6 mm	

[a] particle diameter, pore size, column dimensions

[b] applied in Chapter 8

[c] applied in Chapter 9

11

Concluding Remarks and Perspectives

11.1 Concluding Remarks

A high-temperature one-pot – one-step polymerization method, used to synthesize macromonomers, was demonstrated to be a versatile tool for conducting (follow-up) chemistry based on macromonomers. The straightforward preparation of the vinyl terminated macromolecular building blocks allowed for their facile incorporation into complex macromolecular architecture. Polymeric structures were obtained by utilizing a modular ligation technique for the formation of dendritic structures as well as a (co)polymerization method for the generation of (block) copolymers.

Initially, a fundamental macromonomer library based on acrylate and acrylate-type monomers was established (Chapter 6). The reaction conditions for the high-temperature polymerization process were investigated in order to obtain high purity macromonomers. It was found that macromonomers with a major quantity of unsaturated chain termini were obtained using *n*-butyl acrylate as a monomer. In addition, it was demonstrated that the addition of radical initiators to the macromonomer formation reaction reduces the inhibition phase observed during initiator-free synthesis, without interfering in the actual polymerization. Generally, 82 % to 95 % (as estimated by ESI-MS) of the macromonomer generated in a single reaction possess the geminal double bond, depending on the type of monomer. The molecular weight of the macromonomers that can be achieved is located between 800 $g{\cdot}mol^{-1}$ and 2000 $g{\cdot}mol^{-1}$, with polydispersities close to 1.6.

The initial generated macromonomer library, as described in Chapter 6, was expanded by synthesizing dendronized macromonomeric structures (Chapter 7). Dendronized acrylates based on 2,2-bis(hydroxymethyl)propionic acid (bis-MPA) were synthesized from dendrimers which subsequently were functionalized via *click* chemistry (i.e., copper-catalyzed azide-alkyne cycloaddition (CuAAC)). Utilizing this synthetic procedure, dendronized acrylates from 1st to 3rd generation were generated that contained a 6 or 9 carbon chain spacer between the acrylate functional group and the dendron core. The dendronized macromonomers were obtained by subjecting the dendronized acrylates to the high-temperature polymerization conditions established in Chapter 6. The achieved number-average molecular weight (M_n) of the dendronized macromonomers was between 1700 and 4400 $g \cdot mol^{-1}$. The obtained vinyl terminated oligomers with its geminal substitution were found to have a purity of as high as 83 %, where the 1st generation dendronized acrylates provided the highest purity. Moderate deprotection of the acetonide groups occurred spontaneously during the macromonomer formation process and was most serve for the 3rd generation dendrimers.

In addition to creating dendritic structures (Chapter 7), copolymers were synthesized either via free radical copolymerization (Chapter 8) or by ring-opening polymerization (ROP) (Chapter 9). In both chapters, an *n*-butyl acrylate macromonomer (BAMM) was the starting point of the copolymer synthesis.

The successful copolymerization of BAMM with benzyl acrylate (BzA) as a co-monomer to give statistical copolymers featuring pendant side chains was described (Chapter 7). The obtained pBAMM-*co*-pBzA were found to have a M_n between 8000 $g \cdot mol^{-1}$ and 77000 $g \cdot mol^{-1}$ with a polydispersity of 1.30 to 2.12. The composition of the copolymers was calculated from the ^{1}H-NMR integrals associated with resonances for the BzA aromatic protons and the CH_2 protons of the BAMM. These results were subjected to a (*terminal model*) Mayo-Lewis analysis resulting in estimated reactivity ratios (at ≈ 40 % conversion) of $r_{BzA} = 2.46$ and $r_{BAMM} = 1.79$, indicating a copolymer composition of $F_{BA} < 0.65$ for the copolymers derived from $f_{BzA} > 0.9$ up to a copolymer composition of $F_{BA} > 0.9$ for copolymers with a co-monomer feed of $f_{BzA} < 0.6$.

Next, grafted block copolymers were generated via ROP of ε-caprolactone, initiated by a macroinitiator based on BAMM (Chapter 9). In order to synthesize the macroinitiator, the olefinic termini of the BAMM were transformed into a diol. The OH-terminated macroinitiator $pBA(OH)_2$ was subsequently employed in the ROP process of ε-caprolactone utilizing various catalytic systems (i.e., organo-, metal and enzymatic catalysts). The *in vitro* enzymatic catalysis (Novozym® 435) approach proved to result in the most efficient ROP synthesis. However, side reactions – such as transesterification – occur in each catalytic system and while they cannot be avoided – they can be minimized. The species generated during the enzymatic ROP process include the desired block copolymer pBA-*b*-pCL with grafting on the macromonomer backbone as main species as well as pCL homopolymer and residual macroinitiator.

The chemical structure of the synthesized copolymers was investigated using several analytical techniques, such as electrospray ionization mass spectrometry hyphenated with size exclusion chromatography (SEC/ESI-MS) and liquid chromatography at critical conditions (LCCC).

As has been demonstrated in the present thesis, macromonomers are extremly suitable for constructing well-defined polymer architectures, such as branched, block-, graft-, comb- or star-polymers. In addition to the synthetic procedures that have been explored in the present thesis, several other synthetic strategies can be utilized to synthesize well-defined structures based on the established macromonomer library.

11.2 Perspectives

Modular ligation chemistry represents an efficient method for block copolymer preparation. The transformation of the unsaturated vinyl terminus to potential functionalities was – to some extend – successfully shown for the synthesis of the macroinitiator in Chapter 9; however, the complexicity of follow-up chemistry of the vinyl terminus would potentially be limitless.

In the future, the unsaturated terminus of macromonomers could be employed for conjugation reactions, where theoretically highly efficient block copolymer coupling could be achieved. Possible conjugation methods that could be employed, for instance, are photo- or thermal-induced Diels-Alder chemistry or photo-induced [2+2] cycloadditions.[352-354] An alternative type of coupling reaction that could be applied to macromonomers is metal catalyzed olefin metathesis.[355]

In addition, brush polymers or grafted polymers could conceivably be generated by homo- and copolymerizations of the macromonomers. This polymerizations may be achived by living/controlled free radical polymerization (CRP), including nitroxide-mediated polymerization (NMP),[1] atom transfer radical polymerization (ATRP)[2] and reversible addition fragmentation chain transfer (RAFT) polymerization,[3-4] which allow for excellent control over the molecular weight and polydispersity of the generated macromolecular architectures.

For the future it is a matter of priority to fully exploit vary conjugation chemistry as well as (co)polymerization via CRP processes of the macromonomers for the construction of polymeric architectures.

12

References

(1) Hawker, C. J.; Bosman, A. W.; Harth, E. *Chem. Rev.* **2001**, *101*, 3661-3688.

(2) Matyjaszewski, K.; Xia, J. H. *Chem. Rev.* **2001**, *101*, 2921-2990.

(3) Chiefari, J.; Chong, Y. K.; Ercole, F.; Krstina, J.; Jeffery, J.; Le, T. P. T.; Mayadunne, R. T. A.; Meijs, G. F.; Moad, C. L.; Moad, G.; Rizzardo, E.; Thang, S. H. *Macromolecules* **1998**, *31*, 5559-5562.

(4) Moad, G.; Chong, Y. K.; Postma, A.; Rizzardo, E.; Thang, S. H. *Polymer* **2005**, *46*, 8458-8468.

(5) Willemse, R. X. E.; van Herk, A. M.; Panchenko, E.; Junkers, T.; Buback, M. *Macromolecules* **2005**, *38*, 5098-5103.

(6) Azukizawa, M.; Yamada, B.; Hill, D. J. T.; Pomery, P. J. *Macromol. Chem. Phys.* **2000**, *201*, 774-781.

(7) Yamada, B.; Azukizawa, M.; Yamazoe, H.; Hill, D. J. T.; Pomery, P. J. *Polymer* **2000**, *41*, 5611-5618.

(8) Gilbert, B. C.; Smith, J. R. L.; Milne, E. C.; Whitwood, A. C.; Taylor, P. *J. Chem. Soc., Perkin Trans. 2* **1994**, 1759-1769.

(9) Junkers, T.; Barner-Kowollik, C. *J. Polym. Sci., Part A: Polym. Chem.* **2008**, *46*, 7585-7605.

(10) Chiefari, J.; Jeffery, J.; Mayadunne, R. T. A.; Moad, G.; Rizzardo, E.; Thang, S. H. *Macromolecules* **1999**, *32*, 7700-7702.

(11) Zorn, A.-M.; Junkers, T.; Barner-Kowollik, C. *Macromol. Rapid Commun.* **2009**, *30*, 2028-2035.

(12) Zorn, A.-M.; Malkoch, M.; Carlmark, A.; Barner-Kowollik, C. *Polym. Chem.* **2011**, *2*, 1163-1173.

(13) Zorn, A.-M.; Junkers, T.; Barner-Kowollik, C. *Macromolecules* **2011**, *44*, 6691-6700.

(14) Zorn, A.-M.; Barner-Kowollik, C. *J. Polym. Sci., Part A: Polym. Chem.* **2012**, *50*, 2366-2377.

(15) Barner-Kowollik, C.; Russell, G. T. *Prog. Polym. Sci.* **2009**, *34*, 1211-1259.

(16) Trommsdorff, V. E.; Köhle, H.; Lagally, P. *Macromol. Chem.* **1948**, *1*, 169-198.

(17) Norrish, R. G. W.; Smith, R. R. *Nature* **1942**, *150*, 336-337.

(18) Szwarc, M. *Nature* **1956**, *178*, 1168-1169.

(19) Odian, G., *Principles of Polymerization*. 4 ed.; John Wiley & Sons, Inc.: 2004.

(20) Fischer, H.; Souaille, M. *Macromol. Symp.* **2001**, *174*, 231-240.

(21) Barner-Kowollik, C.; Moad, G., *Handbook of RAFT Polymerization*. Wiley-VCH Weinheim, **2008**.

(22) Hadjichristidis, N.; Pitsikalis, M.; Iatrou, H.; Pispas, S. *Macromol. Rapid Commun.* **2003**, *24*, 979-1013.

(23) Li, D.; Grady, M. C.; Hutchinson, R. A. *Ind. Eng. Chem. Res.* **2005**, *44*, 2506-2517.

(24) Li, D.; Leiza, J. R.; Hutchinson, R. A. *Macromol. Theory Simul.* **2005**, *14*, 554-559.

(25) McHale, R.; Aldabbagh, F.; Carroll, W. M.; Yamada, B. *Macromol. Chem. Phys.* **2005**, *206*, 2054-2066.

(26) Quan, C.; Soroush, M.; Grady, M. C.; Hansen, J. E.; Simonsick, W. J. *Macromolecules* **2005**, *38*, 7619-7628.

(27) Ito, K. *Prog. Polym. Sci.* **1998**, *23*, 581-620.

(28) Ito, K.; Kawaguchi, S. *Adv. Polym. Sci.* **1999**, *142*, 129-178.

(29) Yamada, B.; Zetterlund, P. B.; Sato, E. *Prog. Polym. Sci.* **2006**, *31*, 835-877.

(30) Nikitin, A. N.; Hutchinson, R. A. *Macromolecules* **2005**, *38*, 1581-1590.

(31) Theis, A.; Feldermann, A.; Charton, N.; Davis, T. P.; Stenzel, M. H.; Barner-Kowollik, C. *Polymer* **2005**, *46*, 6797-6809.

(32) Junkers, T.; Theis, A.; Buback, M.; Davis, T. P.; Stenzel, M. H.; Vana, P.; Barner-Kowollik, C. *Macromolecules* **2005**, *38*, 9497-9508.

(33) Scott, G. E.; Senogles, E. *Macromol. Sci. Chem.* **1970**, *4*, 1105-1117.

(34) Scott, G. E.; Senogles, E. *Sci. Rev. Macromol. Chem.* **1973**, *9*, 49-69.

(35) Scott, G. E.; Senogles, E. *Macromol. Sci. Chem.* **1974**, *8*, 753-773.

(36) Wunderlich, W. *Macromol. Chem.* **1976**, *177*, 973-989.

(37) Kaszás, G.; Földes-Berezsnich, T.; Tüdos, F. *Eur. Polym. J.* **1983**, *19*, 469-473.

(38) Madruga, E. L.; Fernández-García, M. *Macromol. Chem. Phys.* **1996**, *197*, 3743-3755.

(39) McKenna, T. F.; Villanueva, A.; Santos, A. M. *J. Polym. Sci., Part A: Polym. Chem.* **1999**, *37*, 571-588.

(40) Fernández-García, M.; Fernández-Sanz, M.; Madruga, E. L. *Macromol. Chem. Phys.* **2000**, *201*, 1840-1845.

(41) Buback, M.; Gilbert, R. G.; Russell, G. T.; Hill, D. J. T.; Moad, G.; O'Driscoll, K. F.; Shen, J.; Winnik, M. A. *J. Polym. Sci., Part A: Polym. Chem.* **1992**, *30*, 851-863.

(42) Olaj, O. F.; Bitai, I.; Hinkelmann, F. *Macromol. Chem.* **1987**, *188*, 1689-1702.

(43) Beuermann, S.; Buback, M. *Prog. Polym. Sci.* **2002**, *27*, 191-254.

(44) Buback, M.; Gilbert, R. G.; Hutchinson, R. A.; Klumperman, B.; Kuchta, F.-D.; Manders, B. G.; O'Driscoll, K. F.; Russell, G. T.; Schweer, J. *Macromol. Chem. Phys.* **1995**, *196*, 3267-3280.

(45) Beuermann, S.; Buback, M.; Davis, T. P.; Gilbert, R. G.; Hutchinson, R. A.; Olaj, O. F.; Russell, G. T.; Schweer, J.; vanHerk, A. M. *Macromol. Chem. Phys.* **1997**, *198*, 1545-1560.

(46) Asua, J. M.; Beuermann, S.; Buback, M.; Castignolles, P.; Charleux, B.; Gilbert, R. G.; Hutchinson, R. A.; Leiza, J. R.; Nikitin, A. N.; Vairon, J.-P.; van Herk, A. M. *Macromol. Chem. Phys.* **2004**, *205*, 2151-2160.

(47) Beuermann, S.; Paquet, D. A.; McMinn, J. H.; Hutchinson, R. A. *Macromolecules* **1996**, *29*, 4206-4215.

(48) Lyons, R. A.; Hutovic, J.; Piton, M. C.; Christie, D. I.; Clay, P. A.; Manders, B. G.; Kable, S. H.; Gilbert, R. G. *Macromolecules* **1996**, *29*, 1918-1927.

(49) Tanaka, K.; Yamada, B.; Willemse, R.; van Herk, A. M. *Polym. J.* **2002**, *34*, 692.

(50) van Herk, A. M. *Macromol. Rapid Commun.* **2001**, *22*, 687-689.

(51) Couvreur, L.; Piteau, G.; Castignolles, P.; Tonge, M.; Coutin, B.; Charleux, B.; Vairon, J.-P. *Macromol. Symp.* **2001**, *174*, 197-208.

(52) Peck, A. N. F.; Hutchinson, R. A. *Macromolecules* **2004**, *37*, 5944-5951.

(53) Plessis, C.; Arzamendi, G.; Alberdi, J. M.; van Herk, A. M.; Leiza, J. R.; Asua, J. M. *Macromol. Rapid Commun.* **2003**, *24*, 173-177.

(54) Arzamendi, G.; Plessis, C.; Leiza, J. R.; Asua, J. M. *Macromol. Theory Simul.* **2003**, *12*, 315-324.

(55) Barner-Kowollik, C.; Günzler, F.; Junkers, T. *Macromolecules* **2008**, *41*, 8971-8973.

(56) Willemse, R. X. E.; van Herk, A. M. *Macromol. Chem. Phys.* **2010**, *211*, 539-545.

(57) Krstina, J.; Moad, C. L.; Moad, G.; Rizzardo, E.; Berge, C. T.; Fryd, M. *Macromol. Symp.* **1996**, *111*, 13-23.

(58) Roedel, M. J. *J. Am. Chem. Soc.* **1953**, *75*, 6110-6112.

(59) Ahmad, N. M.; Heatley, F.; Lovell, P. A. *Macromolecules* **1998**, *31*, 2822-2827.

(60) McCord, E. F.; Shaw, W. H.; Hutchinson, R. A. *Macromolecules* **1997**, *30*, 246-256.

(61) Nikitin, A. N.; Hutchinson, R. A.; Buback, M.; Hesse, P. *Macromolecules* **2007**, *40*, 8631-8641.

(62) Kajiwara, A. *Polym. Prepr. (Am. Chem. Soc., Div. Polym. Chem.)* **2008**, *49*, 104.

(63) van Herk, A. M. *Macromol. Rapid Commun.* **2009**, *30*, 1964-1968.

(64) Junkers, T.; Barner-Kowollik, C. *Macromol. Theory Simul.* **2009**, *18*, 421-433.

(65) Barner-Kowollik, C.; Junkers, T. *J. Polym. Sci., Part A: Polym. Chem.* **2011**, *49*, 1293-1297.

(66) Barner-Kowollik, C.; Davis, T. P.; Stenzel, M. H. *Polymer* **2004**, *45*, 7791-7805.

(67) Junkers, T.; Bennet, F.; Koo, S. P. S.; Barner-Kowollik, C. *J. Polym. Sci., Part A: Polym. Chem.* **2008**, *46*, 3433-3437.

(68) Koo, S. P. S.; Junkers, T.; Barner-Kowollik, C. *Macromolecules* **2009**, *42*, 62-69.

(69) Junkers, T.; Koo, S. P. S.; Davis, T. P.; Stenzel, M. H.; Barner-Kowollik, C. *Macromolecules* **2007**, *40*, 8906-8912.

(70) Chiefari, J.; Moad, G.; Rizzardo, E.; Gridnev, A. A. US 98/47927, **1998**.

(71) Junkers, T.; Koo, S. P. S.; Davis, T. P.; Stenzel, M. H.; Barner-Kowollik, C. *Macromolecules* **2007**, *40*, 8906-8912.

(72) Junkers, T. *Aust. J. Chem.* **2008**, *61*, 646-646.

(73) Enikolopyan, N. S.; Smirnov, B. R.; Ponomarev, G. V.; Belgovskii, I. M. *J. Polym. Sci., Polym. Chem. Ed.* **1981**, *19*, 879-889.

(74) Gridnev, A. *J. Polym. Sci., Part A: Polym. Chem.* **2000**, *38*, 1753-1766.

(75) Cacioli, P.; Moad, G.; Rizzardo, E.; Serelis, A. K.; Solomon, D. H. *Polym. Bull.* **1984**, *11*, 325-328.

(76) Cacioli, P.; Hawthorne, D. G.; Johns, S. R.; Solomon, D. H.; Rizzardo, E.; Willing, R. I. *J. Chem. Soc., Chem. Commun.* **1985**, 1355-1356.

(77) Davis, T. P.; Haddleton, D. M.; Richards, S. N. *Sci. Rev. Macromol. Chem.* **1994**, *34*, 243-324.

(78) Gridnev, A. A.; Ittel, S. D. *Chem. Rev.* **2001**, *101*, 3611-3659.

(79) Heuts, J. P. A.; Roberts, G. E.; Biasutti, J. D. *Aust. J. Chem.* **2002**, *55*, 381-398.

(80) Burczyk, A. F.; O'Driscoll, K. F.; Rempel, G. L. *J. Polym. Sci., Polym. Chem. Ed.* **1984**, *22*, 3255-3262.

(81) Roberts, G. E.; Barner-Kowollik, C.; Davis, T. P.; Heuts, J. P. A. *Macromolecules* **2003**, *36*, 1054-1062.

(82) Pierik, S. C. J.; van Herk, A. M. *Macromol. Chem. Phys.* **2003**, *204*, 1406-1418.

(83) Pierik, B.; Masclee, D.; van Herk, A. *Macromol. Symp.* **2001**, *165*, 19-28.

(84) Roberts, G. E.; Heuts, J. P. A.; Davis, T. P. *J. Polym. Sci., Part A: Polym. Chem.* **2003**, *41*, 752-765.

(85) Gridnev, A. A. *Polym. J.* **1992**, *24*, 613-623.

(86) Sanayei, R. A.; O'Driscoll, K. F. *Macromol. Sci. Chem.* **1989**, *26*, 1137-1149.

(87) Heuts, J. P. A.; Forster, D. J.; Davis, T. P. *Macromol. Rapid Commun.* **1999**, *20*, 299-302.

(88) Heuts, J. P. A.; Smeets, N. M. B. *Polym. Chem.* **2011**, *2*, 2407-2423.

(89) Heuts, J. P. A.; Forster, D. J.; Davis, T. P.; Yamada, B.; Yamazoe, H.; Azukizawa, M. *Macromolecules* **1999**, *32*, 2511-2519.

(90) Suddaby, K. G.; Sanayei, R. A.; Rudin, A.; O'Driscoll, K. F. *J. Appl. Polym. Sci.* **1991**, *43*, 1565-1575.

(91) Yamada, B.; Oku, F.; Harada, T. *J. Polym. Sci., Part A: Polym. Chem.* **2003**, *41*, 645-654.

(92) Chiu, T. Y. J.; Heuts, J. P. A.; Davis, T. P.; Stenzel, M. H.; Barner-Kowollik, C. *Macromol. Chem. Phys.* **2004**, *205*, 752-761.

(93) Soeriyadi, A. H.; Boyer, C.; Burns, J.; Becer, C. R.; Whittaker, M. R.; Haddleton, D. M.; Davis, T. P. *Chem. Commun.* **2010**, *46*, 6338-6340.

(94) Kato, S.; Ishida, M. *Sulfur Rep.* **1988**, *8*, 155-312.

(95) DePuy, C. H.; King, R. W. *Chem. Rev.* **1960**, *60*, 431-457.

(96) Tschugaeff, L. *Chem. Ber.* **1899**, *32*, 3332-3335.

(97) Chong, B.; Moad, G.; Rizzardo, E.; Skidmore, M.; Thang, S. H. *Aust. J. Chem.* **2006**, *59*, 755-762.

(98) Hu, Y.-H.; Chen, C.-Y.; Wang, C.-C. *Polym. Degrad. Stab.* **2004**, *84*, 505-514.

(99) Lehrle, R. S.; Place, E. J. *Polym. Degrad. Stab.* **1997**, *56*, 215-219.

(100) Lehrle, R. S.; Place, E. J. *Polym. Degrad. Stab.* **1997**, *56*, 221-226.

(101) Xu, J.; He, J.; Fan, D.; Tang, W.; Yang, Y. *Macromolecules* **2006**, *39*, 3753-3759.

(102) Zhou, Y.; He, J.; Li, C.; Hong, L.; Yang, Y. *Macromolecules* **2011**, *44*, 8446-8457.

(103) Postma, A.; Davis, T. P.; Moad, G.; O'Shea, M. S. *Macromolecules* **2005**, *38*, 5371-5374.

(104) Postma, A.; Davis, T. P.; Li, G.; Moad, G.; O'Shea, M. S. *Macromolecules* **2006**, *39*, 5307-5318.

(105) Shen, Y.; Zhu, S.; Zeng, F.; Pelton, R. *Macromolecules* **2000**, *33*, 5399-5404.

(106) Shen, Y.; Zhu, S.; Zeng, F.; Pelton, R. *Macromol. Chem. Phys.* **2000**, *201*, 1387-1394.

(107) Öztürk, T.; Hazer, B. *J. Macromol. Sci., Part A: Pure Appl. Chem.* **2010**, *47*, 265-272.

(108) Ferrari, R.; Yu, Y.; Morbidelli, M.; Hutchinson, R. A.; Moscatelli, D. *Macromolecules* **2011**, *44*, 9205-9212.

(109) Schön, F.; Hartenstein, M.; Müller, A. H. E. *Macromolecules* **2001**, *34*, 5394-5397.

(110) Vogt, A. P.; Sumerlin, B. S. *Macromolecules* **2006**, *39*, 5286-5292.

(111) Topham, P. D.; Sandon, N.; Read, E. S.; Madsen, J.; Ryan, A. J.; Armes, S. P. *Macromolecules* **2008**, *41*, 9542-9547.

(112) Falkenhagen, J.; Much, H.; Stauf, W.; Müller, A. H. E. *Macromolecules* **2000**, *33*, 3687-3693.

(113) Moad, G.; Barner-Kowollik, C., The Mechanism and Kinetics of the RAFT Process. In *Handbook of RAFT Polymerization*, Barner-Kowollik, C., Ed. Wiley VCH: Weinheim, **2008**.

(114) Brandrup, J.; H., I. E.; Grulke, E. A., *Polymer Handbook*. 4 ed.; John Wiley & Sons, Inc.: New York, **1991**.

(115) Grubisic, Z.; Rempp, P.; Benoit, H. *J. Polym. Sci., Part B: Polym. Phys.* **1996**, *34*, 1707-1713.

(116) Belenky, B. G.; Gankina, E. S.; Tennikov, M. B.; Vilenchik, L. Z. *J. Chromatogr., A* **1978**, *147*, 99-110.

(117) Chagneux, N.; Trimaille, T.; Rollet, M.; Beaudoin, E.; Gérard, P.; Bertin, D.; Gigmes, D. *Macromolecules* **2009**, *42*, 9435-9442.

(118) Gao, H. F.; Min, K.; Matyjaszewski, K. *Macromol. Chem. Phys.* **2006**, *207*, 1709-1717.

(119) Inglis, A. J.; Barner-Kowollik, C. *Polym. Chem.* **2011**, *2*, 126-136.

(120) Min, K.; Gao, H.; Matyjaszewski, K. *J. Am. Chem. Soc.* **2005**, *127*, 3825-3830.

(121) Pasch, H. *Polymer* **1993**, *34*, 4095-4099.

(122) Pasch, H.; Brinkmann, C.; Gallot, Y. *Polymer* **1993**, *34*, 4100-4104.

(123) Roos, S. G.; Müller, A. H. E.; Matyjaszewski, K. *Controlled/Living Radical Polymerization*, American Chemical Society: **2000**; Vol. 768, 361-371.

(124) Roos, S. G.; Müller, A. H. E.; Matyjaszewski, K. *Macromolecules* **1999**, *32*, 8331-8335.

(125) Roos, S. G.; Schmitt, B.; Müller, A. H. E. *Polym. Prepr.* **1999**, *40*, 984.

(126) Schmid, C.; Falkenhagen, J.; Barner-Kowollik, C. *J. Polym. Sci., Part A: Polym. Chem.* **2011**, *49*, 1-10.

(127) Schmid, C.; Weidner, S.; Falkenhagen, J.; Barner-Kowollik, C. *Macromolecules* **2012**, *45*, 87-99.

(128) Wong, E. H. H.; Stenzel, M. H.; Junkers, T.; Barner-Kowollik, C. *Macromolecules* **2010**, *43*, 3785-3793.

(129) Braun, D.; Esser, E.; Rasch, H. *Int. J. Polym. Anal. Charact* **1998**, *4*, 501-516.

(130) Gancheva, V. B.; Vladimirov, N. G.; Velichkova, R. S. *Macromol. Chem. Phys.* **1996**, *197*, 1757-1770.

(131) Krüger, R. P.; Much, H.; Schulz, G. *J. Liq. Chromatogr.* **1994**, *17*, 3069-3090.

(132) Mengerink, Y.; Peters, R.; deKoster, C. G.; van der Wal, S.; Claessens, H. A.; Cramers, C. A. *J. Chromatogr., A* **2001**, *914*, 131-145.

(133) Li, M.; Jahed, N. M.; Min, K.; Matyjaszewski, K. *Macromolecules* **2004**, *37*, 2434-2441.

(134) Wong, E. H. H.; Stenzel, M. H.; Junkers, T.; Barner-Kowollik, C. *Macromolecules* **2010**, *43*, 3785-3793.

(135) Gruendling, T.; Weidner, S.; Falkenhagen, J.; Barner-Kowollik, C. *Polym. Chem.* **2010**, *1*, 599-617.

(136) Hart-Smith, G.; Barner-Kowollik, C. *Macromol. Chem. Phys.* **2010**, *211*, 1507-1529.

(137) Siuzdak, G. *Proc. Natl. Acad. Sci. U. S. A.* **1994**, *91*, 11290-11297.

(138) Hanton, S. D. *Chem. Rev.* **2001**, *101*, 527-570.

(139) Barner-Kowollik, C.; Gruendling, T.; Falkenhagen, J.; Weidner, S., *Mass Spectrometry in Polymer Chemistry*. Wiley-VCH: Weinheim, **2012**.

(140) McLuckey, S. *J. Am. Soc. Mass Spectrom.* **1992**, *3*, 599-614.

(141) Lehmann, W. D.; Kessler, M.; König, W. A. *Biomed. Mass Spectrom.* **1984**, *11*, 217-222.

(142) Barber, M.; Bordoli, R. S.; Sedgwick, R. D.; Tyler, A. N. *Nature* **1981**, *293*, 270-275.

(143) Barber, M.; Bordoli, R. S.; Elliott, G. J.; Sedgwick, R. D.; Tyler, A. N. *Anal. Chem.* **1982**, *54*, 645A-657A.

(144) Fenn, J. B.; Mann, M.; Meng, C. K.; Wong, S. F.; Whitehouse, C. M. *Mass Spec. Rev.* **1990**, *9*, 37-70.

(145) Hillenkamp, F.; Karas, M.; Beavis, R. C.; Chait, B. T. *Anal. Chem.* **1991**, *63*, 1193A-1203A.

(146) Carroll, D. I.; Dzidic, I.; Horning, E. C.; Stillwell, R. N. *Appl. Spectrosc. Rev.* **1981**, *17*, 337-406.

(147) Horning, E. C.; Horning, M. G.; Carroll, D. I.; Dzidic, I.; Stillwell, R. N. *Anal. Chem.* **1973**, *45*, 936-943.

(148) Byrdwell, W. *Lipids* **2001**, *36*, 327-346.

(149) Vestal, M. L. *Chem. Rev.* **2001**, *101*, 361-376.

(150) Montaudo, G.; Lattimer, R., *Mass Spectrometry of Polymers*. CRC Press: **2002**.

(151) Cole, R. B., *Electrospray and MALDI Mass Spectrometry*. 2 ed.; Wiley VCH: Hoboken, New Jersey, **2010**.

(152) Guittard, J.; Tessier, M.; Blais, J. C.; Bolbach, G.; Rozes, L.; Maréchal, E.; Tabet, J. C. *J. Mass Spectrom.* **1996**, *31*, 1409-1421.

(153) Dole, M.; Mack, L. L.; Hines, R. L.; Mobley, R. C.; Ferguson, L. D.; Alice, M. B. *J. Chem. Phys.* **1968**, *49*, 2240-2249.

(154) Mack, L. L.; Kralik, P.; Rheude, A.; Dole, M. *J. Chem. Phys.* **1970**, *52*, 4977-4986.

(155) Yamashita, M.; Fenn, J. B. *J. Phys. Chem.* **1984**, *88*, 4451-4459.

(156) Yamashita, M.; Fenn, J. B. *J. Phys. Chem.* **1984**, *88*, 4671-4675.

(157) Meng, C. K.; Mann, M.; Fenn, J. B. *Z. Phys. D* **1988**, *10*, 361-368.

(158) Fenn, J.; Mann, M.; Meng, C.; Wong, S.; Whitehouse, C. *Science* **1989**, *246*, 64-71.

(159) Fenn, J. B. *Angew. Chem., Int. Ed.* **2003**, *42*, 3871-3894.

(160) Grace, J. M.; Marijnissen, J. C. M. *J. Aerosol Sci.* **1994**, *25*, 1005-1019.

(161) Gaskell, S. J. *J. Mass Spectrom.* **1997**, *32*, 677-688.

(162) Bruins, A. P. *J. Chromatogr., A* **1998**, *794*, 345-357.

(163) Kebarle, P.; Tang, L. *Anal. Chem.* **1993**, *65*, 972A-986A.

(164) Kebarle, P. *J. Mass Spectrom.* **2000**, *35*, 804-817.

(165) Kebarle, P.; Peschke, M. *Anal Chim Acta* **2000**, *406*, 11-35.

(166) Schmelzeisen-Redeker, G.; Bütfering, L.; Röllgen, F. W. *Int. J. Mass Spectr. Ion Proc.* **1989**, *90*, 139-150.

(167) Iribarne, J. V.; Thomson, B. A. *J. Chem. Phys.* **1976**, *64*, 2287-2294.

(168) Thomson, B. A.; Iribarne, J. V. *J. Chem. Phys.* **1979**, *71*, 4451-4463.

(169) Hogan, C. J.; Carroll, J. A.; Rohrs, H. W.; Biswas, P.; Gross, M. L. *Anal. Chem.* **2008**, *81*, 369-377.

(170) Hogan, C. J.; Carroll, J. A.; Rohrs, H. W.; Biswas, P.; Gross, M. L. *J. Am. Chem. Soc.* **2008**, *130*, 6926-6927.

(171) Karas, M.; Bachmann, D.; Bahr, U.; Hillenkamp, F. *Int. J. Mass Spectr. Ion Proc.* **1987**, *78*, 53-68.

(172) Karas, M.; Bachmann, D.; Hillenkamp, F. *Anal. Chem.* **1985**, *57*, 2935-2939.

(173) Karas, M.; Bahr, U.; Hillenkamp, F. *Int. J. Mass Spectr. Ion Proc.* **1989**, *92*, 231-242.

(174) Dreisewerd, K. *Chem. Rev.* **2003**, *103*, 395-426.

(175) Karas, M.; Bahr, U. *TrAC Trends Anal. Chem.* **1986**, *5*, 90-93.

(176) Liao, P.-C.; Allison, J. *J. Mass Spectrom.* **1995**, *30*, 408-423.

(177) Karas, M.; Krüger, R. *Chem. Rev.* **2003**, *103*, 427-440.

(178) Badman, E. R.; Graham Cooks, R. *J. Mass Spectrom.* **2000**, *35*, 659-671.

(179) Louris, J. N.; Amy, J. W.; Ridley, T. Y.; Cooks, R. G. *Int. J. Mass Spectr. Ion Proc.* **1989**, *88*, 97-111.

(180) March, R. E. *J. Mass Spectrom.* **1997**, *32*, 351-369.

(181) McLuckey, S. A.; Van Berkel, G. J.; Goeringer, D. E.; Glish, G. L. *Anal. Chem.* **1994**, *66*, 689A-696A.

(182) Cooks, R. G.; Kaiser, R. E. *Acc. Chem. Res.* **1990**, *23*, 213-219.

(183) Nielen, M. W. F. *Mass Spec. Rev.* **1999**, *18*, 309-344.

(184) Guilhaus, M. *J. Mass Spectrom.* **1995**, *30*, 1519-1532.

(185) Guilhaus, M.; Mlynski, V.; Selby, D. *Rapid Commun. Mass Spectrom.* **1997**, *11*, 951-962.

(186) Vestal, M. L.; Juhasz, P.; Martin, S. A. *Rapid Commun. Mass Spectrom.* **1995**, *9*, 1044-1050.

(187) Wiley, W. C.; McLaren, I. H. *Rev. Sci. Instrum.* **1955**, *26*, 1150-1157.

(188) Brown, R. S.; Lennon, J. J. *Anal. Chem.* **1995**, *67*, 1998-2003.

(189) Schriemer, D. C.; Li, L. *Anal. Chem.* **1996**, *68*, 2721-2725.

(190) Guilhaus, M.; Selby, D.; Mlynski, V. *Mass Spec. Rev.* **2000**, *19*, 65-107.

(191) Birkinshaw, K. *J. Mass Spectrom.* **1997**, *32*, 795-806.

(192) Busch, K. L. *Spectroscopy* **2000**, *15*, 28-33.

(193) Burroughs, E. G. *Rev. Sci. Instrum.* **1969**, *40*, 35-37.

(194) Allen, J. S. *Phys. Rev.* **1939**, *55*, 966-971.

(195) Allen, J. S. *Rev. Sci. Instrum.* **1947**, *18*, 739-749.

(196) Wiza, J. L. *Nucl. Instrum. Methods* **1979**, *162*, 587-601.

(197) Gruendling, T.; Guilhaus, M.; Barner-Kowollik, C. *Anal. Chem.* **2008**, *80*, 6915-6927.

(198) Cole, R.; Harrata, A. *J. Am. Soc. Mass Spectrom.* **1993**, *4*, 546-556.

(199) Iavarone, A.; Jurchen, J.; Williams, E. *J. Am. Soc. Mass Spectrom.* **2000**, *11*, 976-985.

(200) Barner-Kowollik, C.; Inglis, A. J. *Macromol. Chem. Phys.* **2009**, *210*, 987-992.

(201) Overberger, C. G.; Biletch, H.; Finestone, A. B.; Lilker, J.; Herbert, J. *J. Am. Chem. Soc.* **1953**, *75*, 2078-2082.

(202) van Hook, J. P.; Tobolsky, A. V. *J. Am. Chem. Soc.* **1958**, *80*, 779-782.

(203) Zorn, A.-M. Diploma thesis, University Karlsruhe (TH), Karlsruhe, **2009**.

(204) Müller, M. Dissertation. Göttingen Universty, Göttingen, **2005**.

(205) Grayson, S. M.; Fréchet, J. M. J. *Chem. Rev.* **2001**, *101*, 3819-3867.

(206) Malkoch, M.; Malmström, E. E.; Hult, A. *Macromolecules* **2002**, *35*, 8307-8314.

(207) Buhleier, E.; Wehner, W.; Vogtle, F. *Synthesis* **1978**, 155-158.

(208) Tomalia, D. A.; Fréchet, J. M. J. *J. Polym. Sci., Part A: Polym. Chem.* **2002**, *40*, 2719-2728.

(209) Vögtle, F.; Gestermann, S.; Hesse, R.; Schwierz, H.; Windisch, B. *Prog. Polym. Sci.* **2000**, *25*, 987-1041.

(210) Newkome, G. R.; Yao, Z. Q.; Baker, G. R.; Gupta, V. K. *J. Org. Chem.* **1985**, *50*, 2003-2004.

(211) Newkome, G. R.; Yao, Z. Q.; Baker, G. R.; Gupta, V. K.; Russo, P. S.; Saunders, M. J. *J. Am. Chem. Soc.* **1986**, *108*, 849-850.

(212) Hawker, C.; Fréchet, J. M. J. *J. Chem. Soc., Chem. Commun.* **1990**, 1010-1013.

(213) Hawker, C. J.; Fréchet, J. M. J. *J. Am. Chem. Soc.* **1990**, *112*, 7638-7647.

(214) Inoue, K. *Prog. Polym. Sci.* **2000**, *25*, 453-571.

(215) Carlmark, A.; Hawker, C. J.; Hult, A.; Malkoch, M. *Chem. Soc. Rev.* **2009**, *38*, 352-362.

(216) Fréchet, J. M. J. *Science* **1994**, *263*, 1710-1715.

(217) Lee, C. C.; MacKay, J. A.; Fréchet, J. M. J.; Szoka, F. C. *Nat. Biotech.* **2005**, *23*, 1517-1526.

(218) Matthews, O. A.; Shipway, A. N.; Stoddart, J. F. *Prog. Polym. Sci.* **1998**, *23*, 1-56.

(219) Tomalia, D. A. *Mater. Today* **2005**, *8*, 34-46.

(220) Wooley, K. L.; Hawker, C. J.; Fréchet, J. M. J. *J. Chem. Soc., Perkin Trans. 1* **1991**, 1059-1076.

(221) Hourani, R.; Kakkar, A. *Macromol. Rapid Commun.* **2010**, *31*, 947-974.

(222) Bosman, A. W.; Janssen, H. M.; Meijer, E. W. *Chem. Rev.* **1999**, *99*, 1665-1688.

(223) Esfand, R.; Tomalia, D. A. *Drug Discov. Today* **2001**, *6*, 427-436.

(224) Voit, B. I.; Turner, S. R. *Angew. Macromol. Chem.* **1994**, *223*, 13-27.

(225) Wurm, F.; Frey, H. *Prog. Polym. Sci.* **2011**, *36*, 1-52.

(226) Gao, C.; Yan, D. *Prog. Polym. Sci.* **2004**, *29*, 183-275.

(227) Voit, B. *J. Polym. Sci., Part A: Polym. Chem.* **2000**, *38*, 2505-2525.

(228) Yates, C. R.; Hayes, W. *Eur. Polym. J.* **2004**, *40*, 1257-1281.

(229) Hult, A.; Johannson, M.; Malmström, E. E. *Adv. Polym. Sci.* **1999**, *134*, 1-34.

(230) Fréchet, J. M. J.; Hawker, C. J.; Gitsov, I.; Leon, J. W. *J. Macromol. Sci., Part A: Pure Appl. Chem.* **1996**, *33*, 1399-1425.

(231) Voit, B. I.; Lederer, A. *Chem. Rev.* **2009**, *109*, 5924-5973.

(232) Zhu, X.; Zhou, Y.; Yan, D. *J. Polym. Sci., Part B: Polym. Phys.* **2011**, *49*, 1277-1286.

(233) Konkolewicz, D.; Monteiro, M. J.; Perrier, S. b. *Macromolecules* **2011**, *44*, 7067-7087.

(234) Sunder, A.; Heinemann, J.; Frey, H. *Chem. Eur. J.* **2000**, *6*, 2499-2506.

(235) Gauthier, M. *J. Polym. Sci., Part A: Polym. Chem.* **2007**, *45*, 3803-3810.

(236) Teertstra, S. J.; Gauthier, M. *Prog. Polym. Sci.* **2004**, *29*, 277-327.

(237) Gauthier, M.; Moller, M. *Macromolecules* **1991**, *24*, 4548-4553.

(238) Schlüter, A. D.; Rabe, J. P. *Angew. Chem., Int. Ed.* **2000**, *39*, 864-883.

(239) Frauenrath, H. *Prog. Polym. Sci.* **2005**, *30*, 325-384.

(240) Malkoch, M.; Carlmark, A.; Wodegiorgis, A.; Hult, A.; Malmström, E. E. *Macromolecules* **2004**, *37*, 322-329.

(241) Carlmark, A.; Malmström, E. E. *Macromolecules* **2004**, *37*, 7491-7496.

(242) Helms, B.; Mynar, J. L.; Hawker, C. J.; Fréchet, J. M. J. *J. Am. Chem. Soc.* **2004**, *126*, 15020-15021.

(243) Li, W.; Wu, D. L.; Schlüter, A. D.; Zhang, A. F. *J. Polym. Sci., Part A: Polym. Chem.* **2009**, *47*, 6630-6640.

(244) Hao, X. J.; Malmström, E. E.; Davis, T. P.; Stenzel, M. H.; Barner-Kowollik, C. *Aust. J. Chem.* **2005**, *58*, 483-491.

(245) Xu, J. T.; Tao, L.; Liu, J. Q.; Bulmus, V.; Davis, T. P. *Macromolecules* **2009**, *42*, 6893-6901.

(246) Nystrom, A.; Hult, A. *J. Polym. Sci., Part A: Polym. Chem.* **2005**, *43*, 3852-3867.

(247) Benhabbour, S. R.; Parrott, M. C.; Gratton, S. E. A.; Adronov, A. *Macromolecules* **2007**, *40*, 5678-5688.

(248) Hedrick, J. L.; Trollsas, M.; Hawker, C. J.; Atthoff, B.; Claesson, H.; Heise, A.; Miller, R. D.; Mecerreyes, D.; Jerome, R.; Dubois, P. *Macromolecules* **1998**, *31*, 8691-8705.

(249) Rahm, M.; Westlund, R.; Eldsater, C.; Malmström, E. E. *J. Polym. Sci., Part A: Polym. Chem.* **2009**, *47*, 6191-6200.

(250) Nystrom, A.; Malkoch, M.; Furo, I.; Nystrom, D.; Unal, K.; Antoni, P.; Vamvounis, G.; Hawker, C. J.; Wooley, K.; Malmström, E. E.; Hult, A. *Macromolecules* **2006**, *39*, 7241-7249.

(251) Kolb, H. C.; Finn, M. G.; Sharpless, K. B. *Angew. Chem., Int. Ed.* **2001**, *40*, 2004-2021.

(252) Wu, P.; Feldman, A. K.; Nugent, A. K.; Hawker, C. J.; Scheel, A.; Voit, B.; Pyun, J.; Fréchet, J. M. J.; Sharpless, K. B.; Fokin, V. V. *Angew. Chem., Int. Ed.* **2004**, *43*, 3928-3932.

(253) Joralemon, M. J.; O'Reilly, R. K.; Matson, J. B.; Nugent, A. K.; Hawker, C. J.; Wooley, K. L. *Macromolecules* **2005**, *38*, 5436-5443.

(254) Antoni, P.; Nystrom, D.; Hawker, C. J.; Hult, A.; Malkoch, M. *Chem. Commun.* **2007**, 2249-2251.

(255) Killops, K. L.; Campos, L. M.; Hawker, C. J. *J. Am. Chem. Soc.* **2008**, *130*, 5062-5064.

(256) Montanez, M. I.; Campos, L. M.; Antoni, P.; Hed, Y.; Walter, M. V.; Krull, B. T.; Khan, A.; Hult, A.; Hawker, C. J.; Malkoch, M. *Macromolecules* **2010**, *43*, 6004-6013.

(257) Antoni, P.; Robb, M. J.; Campos, L.; Montanez, M.; Hult, A.; Malmström, E.; Malkoch, M.; Hawker, C. J. *Macromolecules* **2010**, *43*, 6625-6631.

(258) Kolb, H. C.; Sharpless, K. B. *Drug Discov. Today* **2003**, *8*, 1128-1137.

(259) Inglis, A. J.; Stenzel, M. H.; Barner-Kowollik, C. *Macromol. Rapid Commun.* **2009**, *30*, 1792-1798.

(260) Inglis, A. J.; Sinnwell, S.; Stenzel, M. H.; Barner-Kowollik, C. *Angew. Chem., Int. Ed.* **2009**, *48*, 2411-2414.

(261) Wu, P.; Malkoch, M.; Hunt, J. N.; Vestberg, R.; Kaltgrad, E.; Finn, M. G.; Fokin, V. V.; Sharpless, K. B.; Hawker, C. J. *Chem. Commun.* **2005**, 5775-5777.

(262) Malkoch, M.; Schleicher, K.; Drockenmuller, E.; Hawker, C. J.; Russell, T. P.; Wu, P.; Fokin, V. V. *Macromolecules* **2005**, *38*, 3663-3678.

(263) Merz, E.; Alfrey, T.; Goldfinger, G. *J. Polym. Sci.* **1946**, *1*, 75-82.

(264) Alfrey, J. T.; Goldfinger, G. *J. Chem. Phys.* **1944**, *12*, 205-209.

(265) Mayo, F. R.; Lewis, F. M. *J. Am. Chem. Soc.* **1944**, *66*, 1594-1601.

(266) Coote, M. L.; Davis, T. P. *Prog. Polym. Sci.* **1999**, *24*, 1217-1251.

(267) Madruga, E. L. *Prog. Polym. Sci.* **2002**, *27*, 1879-1924.

(268) Fukuda, T.; Kubo, K.; Ma, Y.-D. *Prog. Polym. Sci.* **1992**, *17*, 875-916.

(269) Lewis, F. M.; Walling, C.; Cummings, W.; Briggs, E. R.; Mayo, F. R. *J. Am. Chem. Soc.* **1948**, *70*, 1519-1523.

(270) Nguyen, S.; Marchessault, R. H. *Macromolecules* **2005**, *38*, 290-296.

(271) Davis, T. P. *J. Polym. Sci., Part A: Polym. Chem.* **2001**, *39*, 597-603.

(272) Fukuda, T.; Ma, Y. D.; Inagaki, H. *Macromolecules* **1985**, *18*, 17-26.

(273) Ma, Y. D.; Fukuda, T.; Inagaki, H. *Macromolecules* **1985**, *18*, 26-31.

(274) Davis, T. P.; O'Driscoll, K. F.; Piton, M. C.; Winnik, M. A. *Polym. Int.* **1991**, *24*, 65-70.

(275) Coote, M. L.; Johnston, L. P. M.; Davis, T. P. *Macromolecules* **1997**, *30*, 8191-8204.

(276) Coote, M. L.; Zammit, M. D.; Davis, T. P.; Willett, G. D. *Macromolecules* **1997**, *30*, 8182-8190.

(277) Fukuda, T.; Ma, Y. D.; Inagaki, H.; Kubo, K. *Macromolecules* **1991**, *24*, 370-375.

(278) Fukuda, T.; Goto, A.; Kwak, Y.; Yoshikawa, C.; Ma, Y.-D. *Macromol. Symp.* **2002**, *182*, 53-64.

(279) Fineman, M.; Ross, S. D. *J. Polym. Sci.* **1950**, *5*, 259-262.

(280) Kelen, T.; Tüdos, F. *Macromol. Sci. Chem.* **1975**, *9*, 1-27.

(281) Tidwell, P. W.; Mortimer, G. A. *Sci. Rev. Macromol. Chem.* **1970**, *4*, 281-312.

(282) O'Driscoll, K. F.; Reilly, P. M. *Macromol. Symp.* **1987**, *10-11*, 355-374.

(283) Plaumann, H. P.; Branston, R. E. *J. Polym. Sci., Part A: Polym. Chem.* **1989**, *27*, 2819-2822.

(284) Cai, Y. L.; Hartenstein, M.; Müller, A. H. E. *Macromolecules* **2004**, *37*, 7484-7490.

(285) Krivorotova, T.; Vareikis, A.; Gromadzki, D.; Netopilik, M.; Makuska, R. *Eur. Polym. J.* **2010**, *46*, 546-556.

(286) Norman, J.; Moratti, S. C.; Slark, A. T.; Irvine, D. J.; Jackson, A. T. *Macromolecules* **2002**, *35*, 8954-8961.

(287) Ohno, S.; Matyjaszewski, K. *J. Polym. Sci., Part A: Polym. Chem.* **2006**, *44*, 5454-5467.

(288) Ryan, J.; Aldabbagh, F.; Zetterlund, P. B.; Yamada, B. *React. Funct. Polym.* **2008**, *68*, 692-700.

(289) Meijs, G. F.; Rizzardo, E. *J. Macromol. Sci.-Rev. Macromol. Chem. Phys.* **1990**, *C30*, 305-377.

(290) Penzel, E.; Goetz, N. *Angew. Makromol. Chem.* **1990**, *178*, 191-200.

(291) Wang, W.; Nikitin, A. N.; Hutchinson, R. A. *Macromol. Rapid Commun.* **2009**, *30*, 2022-2027.

(292) Hirano, T.; Zetterlund, P. B.; Yamada, B. *Polym. J.* **2003**, *35*, 491-500.

(293) Chiefari, J.; Chong, Y. K.; Ercole, F.; Krstina, J.; Jeffery, J.; Le, T. P. T.; Mayadunne, R. T. A.; Meijs, G. F.; Moad, C. L.; Moad, G.; Rizzardo, E.; Thang, S. H. *Macromolecules* **1998**, *31*, 5559-5562.

(294) Barner-Kowollik, C. *Handbook of RAFT Polymerization*. Wiley-VCH: Weinheim, **2008**.

(295) Dubois, P.; Coulembier, O.; Raquez, J.-M., *Handbook of Ring-Opening Polymerization*. WILEY-VCH: Weinheim, **2009**.

(296) Kowalski, A.; Duda, A.; Penczek, S. *Macromol. Rapid Commun.* **1998**, *19*, 567-572.

(297) Kowalski, A.; Duda, A.; Penczek, S. *Macromolecules* **2000**, *33*, 7359-7370.

(298) Kricheldorf, H. R.; Kreiser-Saunders, I.; Stricker, A. *Macromolecules* **2000**, *33*, 702-709.

(299) Kricheldorf, H. R. *Macromol. Symp.* **2000**, *153*, 55-65.

(300) von Schenck, H.; Ryner, M.; Albertsson, A.-C.; Svensson, M. *Macromolecules* **2002**, *35*, 1556-1562.

(301) Ryner, M.; Stridsberg, K.; Albertsson, A.-C.; von Schenck, H.; Svensson, M. *Macromolecules* **2001**, *34*, 3877-3881.

(302) Penczek, S.; Duda, A.; Szymanski, R. *Macromol. Symp.* **1998**, *132*, 441-449.

(303) Baran, J.; Duda, A.; Kowalski, A.; Szymanski, R.; Penczek, S. *Macromol. Symp.* **1997**, *123*, 93-101.

(304) Pratt, R. C.; Lohmeijer, B. G. G.; Long, D. A.; Waymouth, R. M.; Hedrick, J. L. *J. Am. Chem. Soc.* **2006**, *128*, 4556-4557.

(305) Lohmeijer, B. G. G.; Pratt, R. C.; Leibfarth, F.; Logan, J. W.; Long, D. A.; Dove, A. P.; Nederberg, F.; Choi, J.; Wade, C.; Waymouth, R. M.; Hedrick, J. L. *Macromolecules* **2006**, *39*, 8574-8583.

(306) Simon, L.; Goodman, J. M. *J. Org. Chem.* **2007**, *72*, 9656-9662.

(307) Meloni, D.; Monaci, R.; Zedde, Z.; Cutrufello, M. G.; Fiorilli, S.; Ferino, I. *Appl. Catal., B* **2011**, *102*, 505-514.

(308) Lakshmi Kantam, M.; Sreekanth, P. *Catal. Lett.* **2001**, *77*, 241-243.

(309) Kobayashi, S.; Makino, A. *Chem. Rev.* **2009**, *109*, 5288-5353.

(310) Kumar, A.; Gross, R. A. *Biomacromolecules* **2000**, *1*, 133-138.

(311) Knani, D.; Gutman, A. L.; Kohn, D. H. *J. Polym. Sci., Part A: Polym. Chem.* **1993**, *31*, 1221-1232.

(312) Kondo, A.; Sugihara, S.; Kuwahara, M.; Toshima, K.; Matsumura, S. *Macromol. Biosci.* **2008**, *8*, 533-539.

(313) Uyama, H.; Kobayashi, S. *Chem. Lett.* **1993**, 1149-1150.

(314) Dong, H.; Cao, S.-G.; Li, Z.-Q.; Han, S.-P.; You, D.-L.; Shen, J.-C. *J. Polym. Sci., Part A: Polym. Chem.* **1999**, *37*, 1265-1275.

(315) Foresti, M. L.; Ferreira, M. L. *Macromol. Rapid Commun.* **2004**, *25*, 2025-2028.

(316) Johnson, P. M.; Kundu, S.; Beers, K. L. *Biomacromolecules* **2011**, *12*, 3337-3343.

(317) Kobayashi, S.; Uyama, H.; Takamoto, T. *Biomacromolecules* **2000**, *1*, 3-5.

(318) Kumar, A.; Gross, R. A. *J. Am. Chem. Soc.* **2000**, *122*, 11767-11770.

(319) Matsumura, S., *Enzyme-Catalyzed Synthesis of Polymers*, Kobayashi, S.; Ritter, H.; Kaplan, D., Eds. Springer-Verlag Berlin **2006**; Vol. 194, 95-132.

(320) Namekawa, S.; Suda, S.; Uyama, H.; Kobayashi, S. *Int. J. Biol. Macromol.* **1999**, *25*, 145-151.

(321) Panova, A. A.; Kaplan, D. L. *Biotechnol. Bioeng.* **2003**, *84*, 103-113.

(322) Kobayashi, S.; Uyama, H.; Namekawa, S. *Polym. Degrad. Stab.* **1998**, *59*, 195-201.

(323) Uyama, H.; Takeya, K.; Kobayashi, S. *Bull. Chem. Soc. Jpn.* **1995**, *68*, 56-61.

(324) Cordova, A.; Iversen, T.; Hult, K. *Polymer* **1999**, *40*, 6709-6721.

(325) Cordova, A.; Iversen, T.; Martinelle, M. *Polymer* **1998**, *39*, 6519-6524.

(326) Mei, Y.; Kumar, A.; Gross, R. *Macromolecules* **2003**, *36*, 5530-5536.

(327) Gandini, A. *Macromolecules* **2008**, *41*, 9491-9504.

(328) Mecking, S. *Angew. Chem., Int. Ed.* **2004**, *43*, 1078-1085.

(329) Meier, M. A. R.; Metzger, J. O.; Schubert, U. S. *Chem. Soc. Rev.* **2007**, *36*, 1788-1802.

(330) Williams, C. K. *Chem. Soc. Rev.* **2007**, *36*, 1573-1580.

(331) Jérôme, C.; Lecomte, P. *Adv. Drug Delivery Rev.* **2008**, *60*, 1056-1076.

(332) Seyednejad, H.; Ghassemi, A. H.; van Nostrum, C. F.; Vermonden, T.; Hennink, W. E. *J. Controlled Release* **2011**, *152*, 168-176.

(333) Zimmer, R.; Homann, K.; Angermann, J.; Reissig, H. U. *Synthesis* **1999**, 1223-1235.

(334) Dietrich, M.; Glassner, M.; Gruendling, T.; Schmid, C.; Falkenhagen, J.; Barner-Kowollik, C. *Polym. Chem.* **2010**, *1*, 634-644.

(335) Xu, J. T.; He, J. P.; Fan, D. Q.; Wang, X. J.; Yang, Y. L. *Macromolecules* **2006**, *39*, 8616-8624.

(336) Kamber, N. E.; Jeong, W.; Waymouth, R. M.; Pratt, R. C.; Lohmeijer, B. G. G.; Hedrick, J. L. *Chem. Rev.* **2007**, *107*, 5813-5840.

(337) Albertsson, A.-C.; Varma, I. K. *Biomacromolecules* **2003**, *4*, 1466-1486.

(338) Chen, B.; Hu, J.; Miller, E. M.; Xie, W.; Cai, M.; Gross, R. A. *Biomacromolecules* **2008**, *9*, 463-471.

(339) Sivalingam, G.; Madras, G. *Biomacromolecules* **2004**, *5*, 603-609.

(340) Dechy-Cabaret, O.; Martin-Vaca, B.; Bourissou, D. *Chem. Rev.* **2004**, *104*, 6147-6176.

(341) Kricheldorf, H. R.; Kreiser-Saunders, I.; Damrau, D.-O. *Macromol. Symp.* **2000**, *159*, 247-258.

(342) Schuchardt, U.; Vargas, R. M.; Gelbard, G. *J. Mol. Catal. A: Chem.* **1995**, *99*, 65-70.

(343) Takwa, M.; Xiao, Y.; Simpson, N.; Malmström, E. E.; Hult, K.; Koning, C. E.; Heise, A.; Martinelle, M. *Biomacromolecules* **2008**, *9*, 704-710.

(344) de Geus, M.; Peeters, J.; Wolffs, M.; Hermans, T.; Palmans, A. R. A.; Koning, C. E.; Heise, A. *Macromolecules* **2005**, *38*, 4220-4225.

(345) Guillarme, D.; Heinisch, S. *Sep. Purif. Rev.* **2005**, *34*, 181-216.

(346) Megoulas, N. C.; Koupparis, M. A. *Crit. Rev. Anal. Chem.* **2005**, *35*, 301-316.

(347) Strazielle, C.; Benoit, H.; Vogl, O. *Eur. Polym. J.* **1978**, *14*, 331-334.

(348) Dervaux, B.; Junkers, T.; Schneider-Baumann, M.; Du Prez, F. E.; Barner-Kowollik, C. *J. Polym. Sci., Part A: Polym. Chem.* **2009**, *47*, 6641-6654.

(349) Gruendling, T.; Junkers, T.; Guilhaus, M.; Barner-Kowollik, C. *Macromol. Chem. Phys.* **2010**, *211*, 520-528.

(350) Schindler, A.; Hibionada, Y. M.; Pitt, C. G. *J. Polym. Sci., Part A: Polym. Chem.* **1982**, *20*, 319-326.

(351) Gruendling, T.; Guilhaus, M.; Barner-Kowollik, C. *Anal. Chem.* **2008**, *80*, 6915-6927.

(352) Glassner, M.; Oehlenschlaeger, K. K.; Gruendling, T.; Barner-Kowollik, C. *Macromolecules* **2011**, *44*, 4681-4689.

(353) Gruendling, T.; Kaupp, M.; Blinco, J. P.; Barner-Kowollik, C. *Macromolecules* **2010**, *44*, 166-174.

(354) Gruendling, T.; Oehlenschlaeger, K. K.; Frick, E.; Glassner, M.; Schmid, C.; Barner-Kowollik, C. *Macromol. Rapid Commun.* **2011**, *32*, 807-812.

(355) Grubbs, R. H.; Chang, S. *Tetrahedron* **1998**, *54*, 4413-4450.

13

Appendix

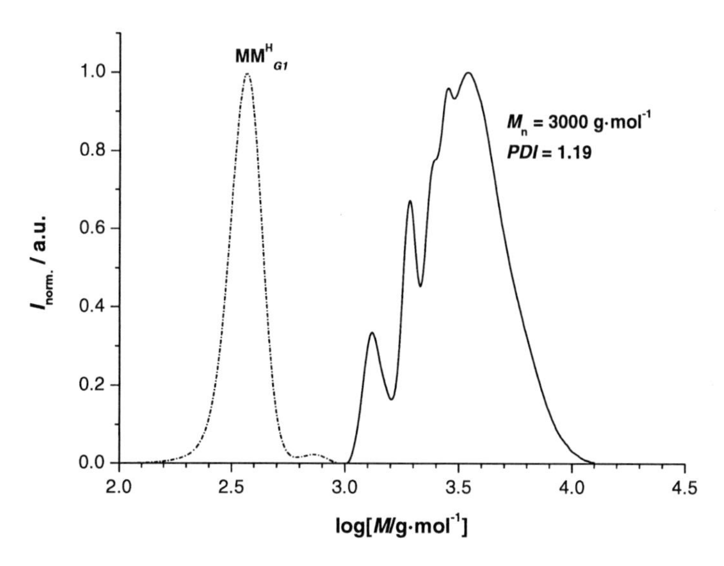

Figure 13-1. SEC chromatogram of p(**11a**) MM. The trace on the left hand side (dashed) represents the residual monomer, whereas the trace on the right hand side shows the distribution after the polymerization. Both traces have been analyzed and normalized separately.

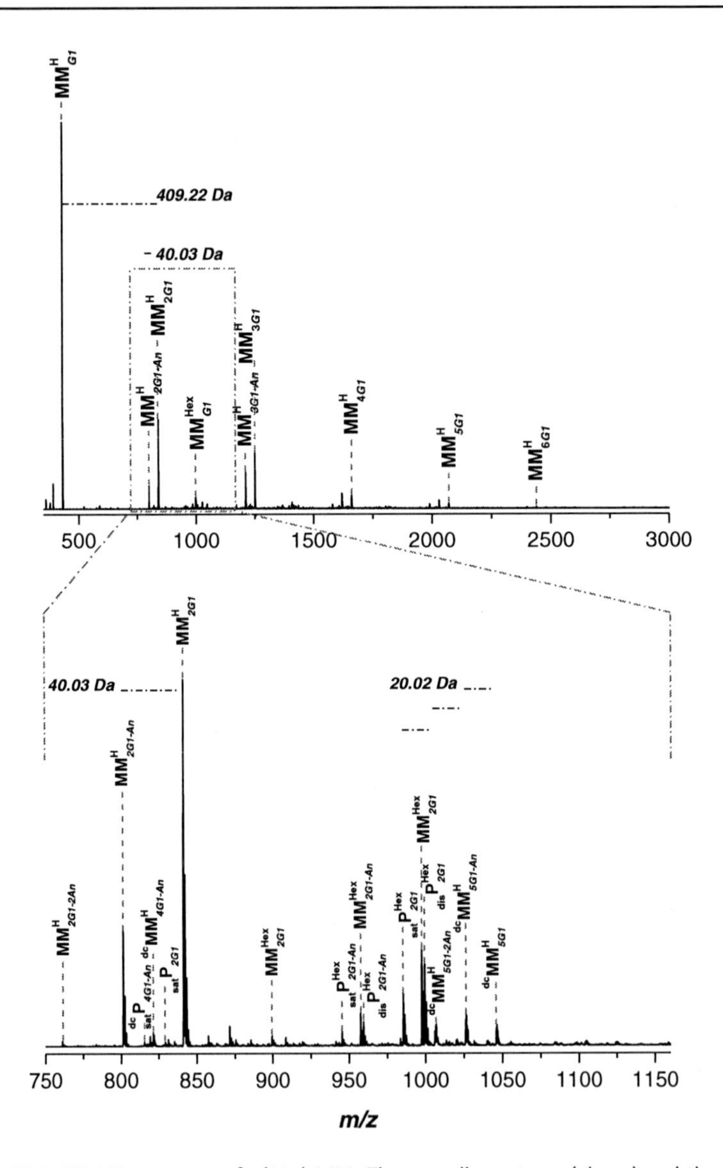

Figure 13-2. ESI-MS spectrum of p(**11a**) MM. The overall spectrum (above) and the zoom spectrum to a repeat unit (409.22 Da) (below) are depicted. The polymer was synthesized via high-temperature acrylate polymerization in solution of hexyl acetate with 5 wt% monomer at 140 °C in an oxygen free atmosphere with 5×10^{-3} mol·L^{-1} AIBN.

Table 13-1. Theoretical and experimental *m/z* ratios of the species of the p(**11a**) MM identified via the ESI-MS measurements. The resolution is close to 0.1 amu (see Figure 13-2). Listed below are the single charged sodium species $[M+Na]^+$ and the double charged sodium species $[M+2Na]^{2+}$ (dc).

species	G1	- An	$m/z_{theo.}$	$m/z_{exp.}$	$\Delta\, m/z$
MM^H	2	2	761.37	761.36	0.01
MM^H	2	1	779.42	779.24	0.18
MM^H dc	4	3	781.39	781.40	0.01
$_{sat}P$	2	1	789.40	789.40	0.00
$_{sat}P$ dc	4	2	795.40	795.40	0.00
MM^H	2	1	801.40	801.40	0.00
MM^H dc	4	2	801.40	801.40	0.00
$_{sat}P$ dc	4	1	815.42	815.40	0.02
MM^H	2	1	817.37	817.40	0.03
MM^H dc	4	2	817.37	817.40	0.03
MM^H	2	0	819.45	819.20	0.25
MM^H dc	4	1	821.42	821.36	0.06
$_{sat}P$	2	0	829.43	829.40	0.03
$_{sat}P$ dc	4	0	835.43	835.36	0.07
MM^H	2	0	841.43	841.40	0.03
MM^H dc	4	0	841.43	841.40	0.03
MM^H	2	0	857.41	857.40	0.01
MM^{Hex} dc	4	3	859.44	859.40	0.04
$_{sat}P^{Hex}$ dc	4	2	873.46	873.40	0.06
MM^{Hex} dc	4	2	879.46	879.40	0.06
MM^{Hex} dc	4	1	899.47	899.44	0.03
$_{sat}P^{Hex}$	2	2	905.48	905.40	0.08
$satP^{Hex}$ dc	4	0	913.49	913.44	0.05
MM^{Hex}	2	2	917.48	917.40	0.08
MM^{Hex} dc	4	0	919.49	919.40	0.09
$_{dis}P^{Hex}$ dc	4	0	920.50	920.40	0.10
$_{sat}P$ dc	5	5	939.96	939.44	0.52

species	G1	- An	m/z theo.	m/z exp.	Δ m/z
$_{sat}P^{Hex}$	1	1	945.52	945.44	0.08
MM^{H} dc	5	5	945.96	945.96	0.00
MM^{Hex}	1	1	957.52	957.44	0.08
$_{dis}P^{Hex}$	1	1	959.98	959.53	0.45
$_{sat}P$ dc	5	3	980.00	980.44	0.44
$_{vic}P^{Hex}$	2	0	983.53	983.48	0.05
$_{sat}P^{Hex}$	2	0	985.55	985.48	0.07
MM^{H} dc	5	3	986.00	985.92	0.08
MM^{Hex}	2	0	997.55	997.48	0.07
$_{dis}P^{Hex}$	2	0	999.56	999.44	0.12
MM^{H} dc	5	2	1006.01	1005.96	0.05
$_{sat}P$ dc	5	1	1020.03	1019.96	0.07
MM^{H} dc	5	1	1026.03	1025.92	0.11
$_{sat}P$ dc	5	0	1040.04	1039.96	0.08
MM^{H} dc	5	0	1046.04	1045.96	0.08
MM^{Hex} dc	5	3	1064.05	1064.48	0.43
$_{sat}P^{Hex}$ dc	5	2	1078.07	1077.96	0.11
MM^{Hex} dc	3	2	1084.07	1083.96	0.11
$_{vic}P^{Hex}$ dc	5	1	1097.08	1097.48	0.40
$_{sat}P^{Hex}$ dc	5	1	1098.08	1098.48	0.40
MM^{Hex} dc	5	1	1104.08	1104.00	0.08
$_{sat}P^{Hex}$ dc	5	0	1118.10	1118.10	0.00
MM^{Hex} dc	5	0	1124.10	1124.00	0.10

Table 13-2. Theoretical and experimental m/z ratios of the species of p(**11a**)-*co*-EA MM identified via the ESI-MS measurements. The resolution is close to 0.1 amu (see Figure 7-14). Listed below are the single charged sodium species $[M+Na]^+$ if not otherwise specified.

species	G1	-An	EA	$m/z_{\text{theo.}}$	$m/z_{\text{exp.}}$	$\Delta m/z$
MM^{Hex}	2	2	4	1317.69	1317.60	0.09
$_{sat}P$	3	3	2	1318.66	1318.60	0.06
$_{sat}P$	1	0	9	1320.68	1320.60	0.08
MM^{H}	0	0	13	1323.67	1323.68	0.01
MM^{Hex}	3	2	0	1326.71	1326.60	0.11
$_{sat}P$	2	0	5	1329.69	1329.60	0.09
MM^{H}	3	3	2	1330.66	1330.60	0.06
MM^{H}	1	0	9	1332.68	1332.60	0.08
$_{sat}P^{Hex}$	1	1	8	1336.71	1336.60	0.11
$_{sat}P$	3	0	1	1338.71	1338.64	0.07
MM^{H}	2	0	5	1341.69	1341.60	0.09
$_{sat}P^{Hex}$	2	1	4	1345.73	1345.60	0.13
MM^{Hex}	1	1	8	1348.71	1348.60	0.11
$_{sat}P$	2	2	6	1349.68	1349.60	0.08
MM^{H}	3	0	1	1350.71	1350.60	0.11
$_{sat}P^{Hex}$	3	1	0	1354.74	1354.64	0.10
MM^{Hex}	2	1	4	1357.73	1357.60	0.13
$_{sat}P$	3	2	2	1358.70	1358.60	0.10
MM^{H}	2	2	6	1361.68	1361.60	0.08
MM^{Hex}	2	0	1	1366.74	1366.60	0.14
$_{sat}P^{Hex}$	0	0	12	1367.73	1367.60	0.13
MM^{H}	3	2	2	1370.70	1370.60	0.10
$_{sat}P^{Hex}$	3	3	1	1374.73	1374.64	0.09
$_{sat}P^{Hex}$	1	0	8	1376.75	1376.64	0.11
MM^{Hex}	0	0	12	1379.73	1379.68	0.05
$_{sat}P$	1	1	10	1380.70	1380.60	0.10
$_{sat}P^{Hex}$	2	0	0	1385.76	1385.64	0.12
MM^{Hex}	3	0	1	1386.73	1386.64	0.09

species	G1	-An	EA	$m/z_{theo.}$	$m/z_{exp.}$	$\Delta m/z$
MM^{Hex}	1	0	8	1388.75	1388.64	0.11
$_{sat}P$	2	1	6	1389.72	1389.64	0.08
MM^{H}	1	1	10	1392.70	1392.60	0.10
$_{sat}P^{Hex}$	3	0	0	1394.77	1394.60	0.17
MM^{Hex}	2	0	4	1397.76	1397.64	0.12
$_{sat}P$	3	1	2	1398.73	1398.64	0.09
MM^{H}	2	1	6	1401.72	1401.60	0.12
$_{sat}P$	2	2	5	1405.75	1405.64	0.11
MM^{Hex}	3	0	0	1406.77	1406.64	0.13
MM^{H}	3	1	2	1410.73	1410.60	0.13
$_{sat}P$	0	0	14	1411.72	1411.64	0.08

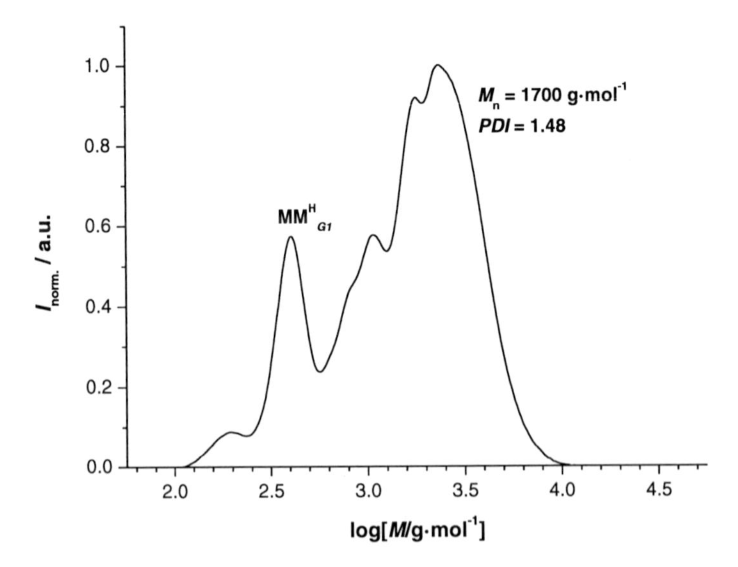

Figure 13-3. SEC chromatogram of p(**11b**)-*co*-EA MM. The peak on the left hand side represents the residual monomer whereas the shifted trace on the right hand side shows the distribution of the copolymerization product.

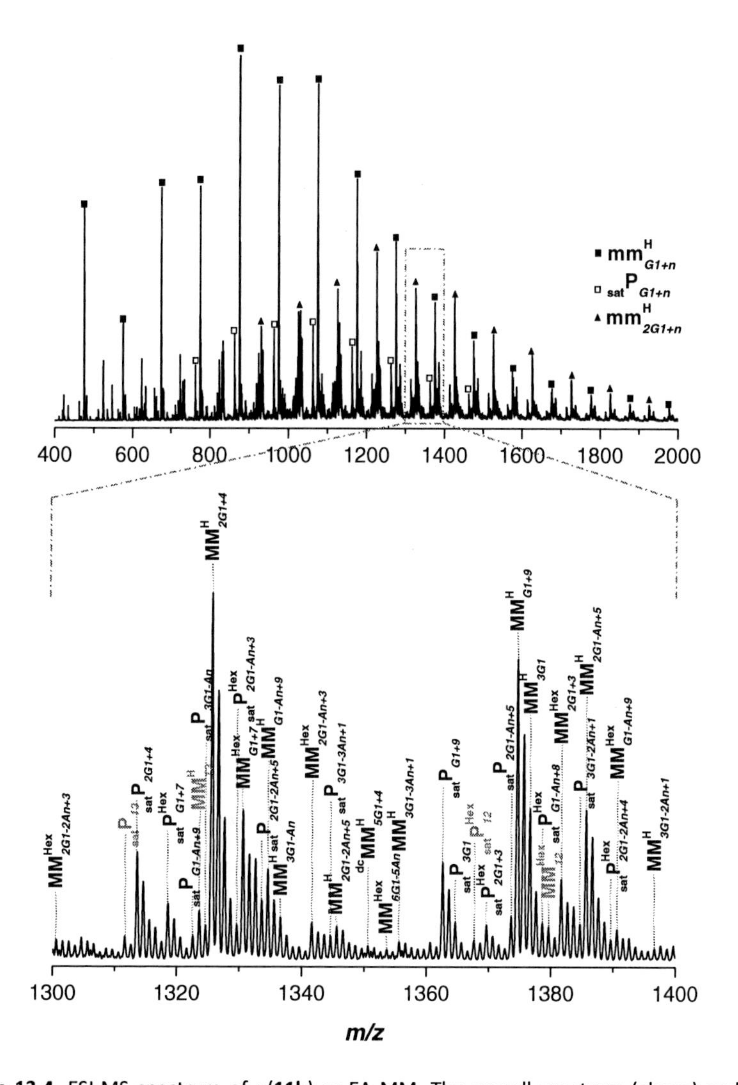

Figure 13-4. ESI-MS spectrum of p(**11b**)-*co*-EA MM. The overall spectrum (above) and the zoom spectrum to a repeat unit (100.05 Da EA) (below) are depicted. The polymer was synthesized via high-temperature acrylate polymerization in solution of hexyl acetate with 2.5 wt% monomer and 2.5 wt% EA at 140 °C in an oxygen free atmosphere with 5×10^{-3} mol·L^{-1} AIBN.

Table 13-3. Theoretical and experimental *m/z* ratios of the species of p(**11b**)-*co*-EA MM identified via the ESI-MS measurements. The resolution is close to 0.1 amu (see Figure 13-4). Listed below are the single charged sodium species $[M+Na]^+$ if not otherwise specified.

species	G1	G1-An	EA	$m/z_{theo.}$	$m/z_{exp.}$	$\Delta m/z$
MM^{Hex}	2	2	4	1317.69	1317.60	0.09
$_{sat}P$	3	3	2	1318.66	1318.60	0.06
$_{sat}P$	1	0	9	1320.68	1320.60	0.08
MM^{H}	0	0	13	1323.67	1323.68	0.01
MM^{Hex}	3	2	0	1326.71	1326.60	0.11
$_{sat}P$	2	0	5	1329.69	1329.60	0.09
MM^{H}	3	3	2	1330.66	1330.60	0.06
MM^{H}	1	0	9	1332.68	1332.60	0.08
$_{sat}P^{Hex}$	1	1	8	1336.71	1336.60	0.11
$_{sat}P$	3	0	1	1338.71	1338.64	0.07
MM^{H}	2	0	5	1341.69	1341.60	0.09
$_{sat}P^{Hex}$	2	1	4	1345.73	1345.60	0.13
MM^{Hex}	1	1	8	1348.71	1348.60	0.11
$_{sat}P$	2	2	6	1349.68	1349.60	0.08
MM^{H}	3	0	1	1350.71	1350.60	0.11
$_{sat}P^{Hex}$	3	1	0	1354.74	1354.64	0.10
MM^{Hex}	2	1	4	1357.73	1357.60	0.13
$_{sat}P$	3	2	2	1358.70	1358.60	0.10
MM^{H}	2	2	6	1361.68	1361.60	0.08
MM^{Hex}	2	0	1	1366.74	1366.60	0.14
$_{sat}P^{Hex}$	0	0	12	1367.73	1367.60	0.13
MM^{H}	3	2	2	1370.70	1370.60	0.10
$_{sat}P^{Hex}$	3	3	1	1374.73	1374.64	0.09
$_{sat}P^{Hex}$	1	0	8	1376.75	1376.64	0.11
MM^{Hex}	0	0	12	1379.73	1379.68	0.05
$_{sat}P$	1	1	10	1380.70	1380.60	0.10
$_{sat}P^{Hex}$	2	0	0	1385.76	1385.64	0.12
MM^{Hex}	3	0	1	1386.73	1386.64	0.09

species	G1	G1-An	EA	$m/z_{\text{theo.}}$	$m/z_{\text{exp.}}$	$\Delta\, m/z$
MM^{Hex}	1	0	8	1388.75	1388.64	0.11
$_{sat}P$	2	1	6	1389.72	1389.64	0.08
MM^{H}	1	1	10	1392.70	1392.60	0.10
$_{sat}P^{Hex}$	3	0	0	1394.77	1394.60	0.17
MM^{Hex}	2	0	4	1397.76	1397.64	0.12
$_{sat}P$	3	1	2	1398.73	1398.64	0.09
MM^{H}	2	1	6	1401.72	1401.60	0.12
$_{sat}P$	2	2	5	1405.75	1405.64	0.11
MM^{Hex}	3	0	0	1406.77	1406.64	0.13
MM^{H}	3	1	2	1410.73	1410.60	0.13
$_{sat}P$	0	0	14	1411.72	1411.64	0.08

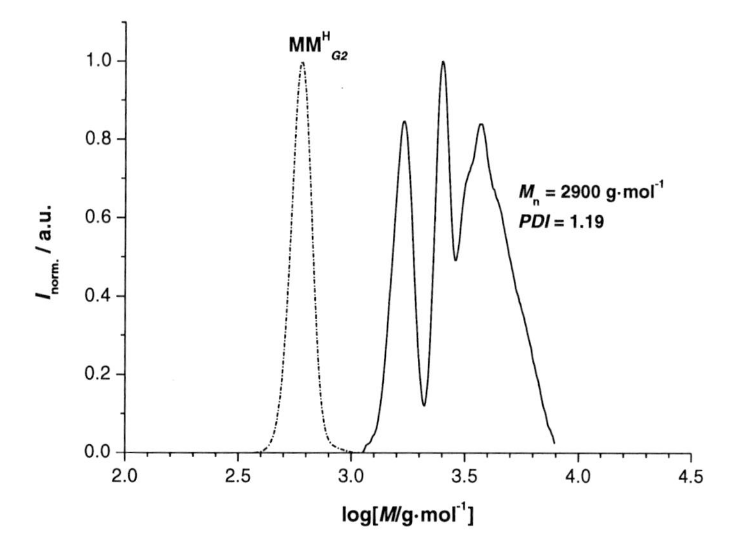

Figure 13-5. SEC chromatogram of p(**12a**) MM. The trace on the left hand side (dashed) represents the residual monomer whereas the trace on the right hand side shows the distribution after the polymerization. Both traces have been analyzed and normalized separately.

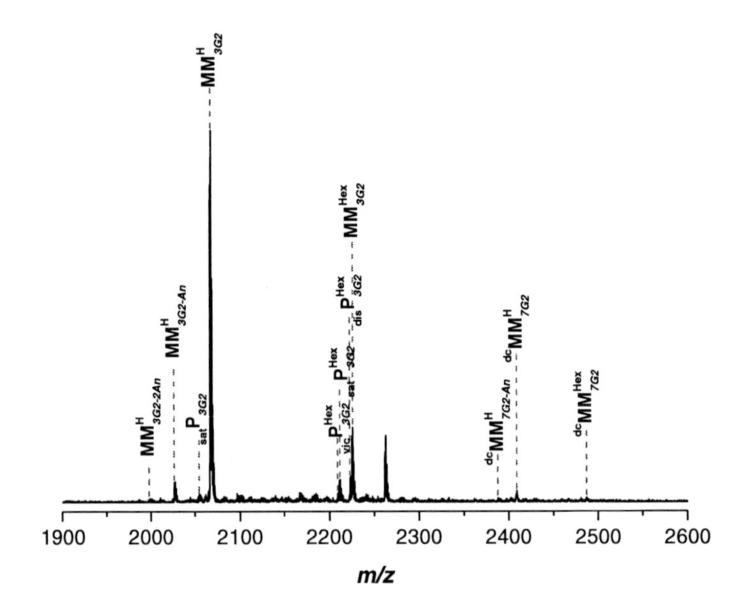

Figure 13-6. ESI-MS spectrum of p(**12a**) MM. The zoom spectrum to a repeat unit (681.35 Da) is depicted. The polymer was synthesized via high-temperature acrylate polymerization in solution of hexyl acetate with 5 wt% monomer at 140 °C in an oxygen free atmosphere with 5×10^{-3} mol·L^{-1} AIBN.

Table 13-4. Theoretical and experimental *m/z* ratios of the species of p(**12a**) MM identified via the ESI-MS measurements. The resolution is close to 0.1 amu (see Figure 13-6). Listed below are the single charged sodium species $[M+Na]^+$ if not otherwise specified.

species	G3	- An	$m/z_{theo.}$	$m/z_{exp.}$	$\Delta m/z$
MM^H	3	2	1986.97	1986.82	0.15
MM^H	3	1	2027.00	2026.82	0.18
$_{sat}P$	3	0	2055.03	2055.00	0.03
MM^H	3	0	2067.03	2066.91	0.12
MM^H $[K]^+$	3	0	2083.01	2083.00	0.01
$_{vic}P^{Hex}$	3	0	2209.13	2209.09	0.04
$_{sat}P^{Hex}$	3	0	2211.15	2210.91	0.24
MM^{Hex}	3	0	2223.15	2223.00	0.15
$_{dis}P^{Hex}$	3	0	2225.16	2225.09	0.07
unassigned				2261.91	
$_{sat}P$ dc	7	1	2383.16	2382.91	0.25
MM^H dc	7	1	2389.16	2389.09	0.07
$_{sat}P$ dc	7	0	2403.19	2403.18	0.01
MM^H dc	7	0	2409.19	2409.09	0.10
$_{sat}P^{Hex}$ dc	7	0	2481.30	2481.00	0.30
MM^{Hex} dc	7	0	2487.30	2487.09	0.21

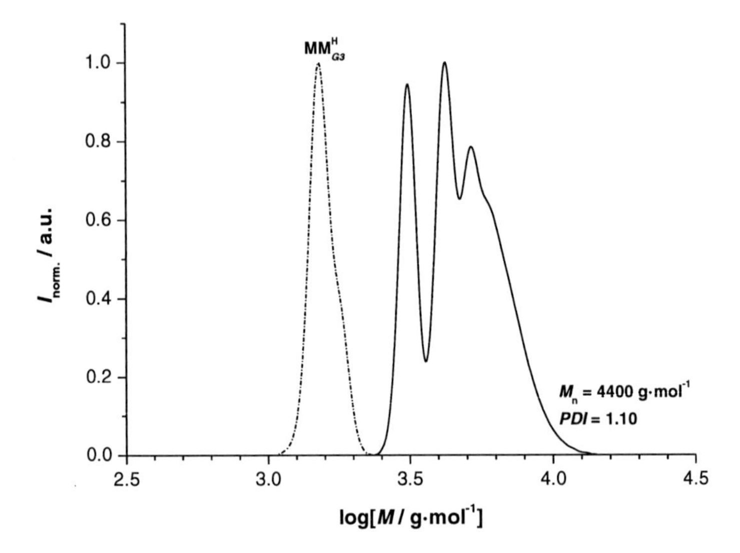

Figure 13-7. SEC chromatogram of p(**13b**) MM. The trace on the left hand side (dashed) represents the residual monomer whereas the trace on the right hand side shows the distribution after the polymerization. Both traces have been analyzed and normalized separately to focus on the formation of species with higher molecular weight.

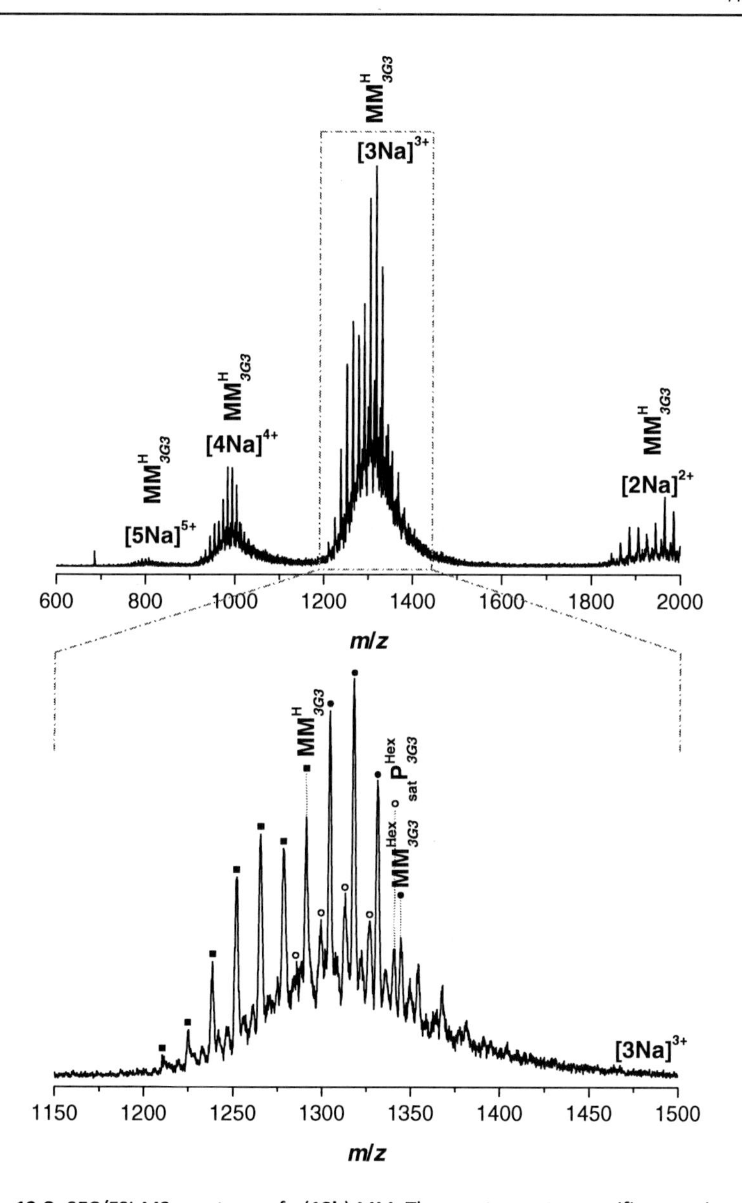

Figure 13-8. SEC/ESI-MS spectrum of p(**13b**) MM. The spectrum at a specific retention time (14.53-14.80 min) is shown above, whereas the spectrum below represents the zoom to the detected triple charged species. The polymer was synthesized via high-temperature acrylate polymerization in solution of hexyl acetate with 5 wt% monomer at 140 °C in an oxygen free atmosphere with 5×10^{-3} mol·L^{-1} AIBN.

Table 13-5. Theoretical and experimental *m/z* ratios of the species of p(**13b**) MM identified via the ESI-MS measurements (see Figure 13-8). The resolution is close to 0.1 amu. Listed below are the triple charged sodium species $[M+3Na]^{3+}$ if not otherwise specified. For the calculation the molar mass was used due to less resolution at higher charges.

species	G3	- An	$m/z_{theo.}$	$m/z_{exp.}$	$\Delta m/z$
MM^{H}	3	6	1211.30	1211.58	0.28
MM^{H}	3	5	1224.66	1225.00	0.34
MM^{H}	3	4	1238.01	1238.58	0.57[a]
MM^{H}	3	3	1251.37	1252.00	0.63[a]
MM^{H}	3	2	1264.72	1265.25	0.53[a]
$_{sat}P^{Hex}$	3	5	1272.73	1272.00	0.73[a]
MM^{H}	3	1	1278.07	1278.42	0.35
$_{sat}P^{Hex}$	3	4	1286.08	1285.75	0.33
MM^{H}	3	0	1291.43	1291.00	0.43
$_{sat}P^{Hex}$	3	3	1299.44	1299.33	0.11
MM^{Hex}	3	3	1303.44	1304.17	0.73[a]
$_{sat}P^{Hex}$	3	2	1312.79	1313.00	0.21
MM^{Hex}	3	2	1316.79	1317.58	0.79[a]
$_{sat}P^{Hex}$	3	1	1326.15	1326.58	0.43
MM^{Hex}	3	1	1330.15	1330.92	0.77[a]
$_{sat}P^{Hex}$	3	0	1339.50	1340.33	0.83[a]
MM^{Hex}	3	0	1343.50	1344.58	1.08[a]

[a] assignments outside experimental accuracy

List of Abbreviations

2-EHA	2-hydroxyethyl acrylate
A	frequency factor (exponential factor in the Arrhenius equation)
AIBN	1,1'-azobis(isobutyronitrile)
amu	atomic mass unit
ATRP	atom transfer radical polymerization
BA	*n*-butyl acrylate
BAMM	*n*-butyl acrylate macromonomer
BECA	2-[[(butylamino)carbonyl]oxy]ethyl acrylate
bis-MPA	2,2-bis(hydroxymethyl)propionic monomer
BzA	benzyl acrylate
CALB	*candida antarctica* lipase B
CCT	Catalytic chain transfer
COBF	Co(II) cobaloxime complex
CRP	controlled radical polymerization
CuAAC	Copper-catalyzed azide-alkyne cycloaddition
Da	Dalton
DC	direct current
dc	double charged
DCM	dichlormethane
DMAP	4-(dimethylamino)-pyridine
DMSO	dimethylsulfoxide
DoPAT	2-[(dodecylsulfanyl)carbonothioyl]sulfanyl propanoic acid
DP_n	degree of polymerization
EA	ethyl acrylate
E_A	activation energy (in the Arrhenius equation)
EAM	enzyme activated monomer
ELSD	evaporative light scattering detector

eROP	enzyme catalyzed ROP
ESI	electrospray ionization
ESI-MS	electrospray ionization mass spectrometry
ESR	electron spin resonance
f	initiation efficiency
FRP	free radical polymerization
HPLC	high pressure liquid chromatography
iBoA	isoborny acrylate
IUPAC	international union of pure and applied chemistry
k_d	rate coefficient of decomposition
k_p^i	rate coefficient of propagation
k_{rei}	rate coefficient of reinitiation
$k_t^{i,j}$	rate coefficient of termination
k_{tr}	rate coefficient of transfer
LAC	liquid adsorption chromatography
LCCC	liquid chromatography at critical conditions
m/z	mass-to-charge
MALDI	matrix-assisted laser desorption ionization
MCR	mid-chain radical
MeOH	methanol
MM	macromonomer
MMA	methyl methacrylate
MM^H	macromonomer with an unsaturated terminus
MM^X	macromonomer with an unsaturated terminus initiated by fragment X
M_n	number-average molecular weight
MS	mass spectrometry
M_w	weight-average molecular weight
NMP	nitroxide-mediated polymerization
NMR	nuclear magnetic resonance
PAMAM	poly(amidoamine)
pBA	poly(*n*-butyl acrylate)
pCL	poly(ε-caprolactone)
PDI	polydispersity index

PLP	pulsed laser polymerization
ppm	parts per million
PRE	persistent radical effect
RAFT	reversible addition fragmentation chain transfer
rf	radio frequency
RI	refractive index
ROP	ring-opening polymerization
R_P	rate of polymerization
rt	retention time
$_{sat}P$	saturated species
SEC	size exclusion chromatography
SEC/ESI-MS	ESI-MS hyphenated with SEC
SFRP	stable free radical polymerization
SPR	secondary propagating radical
***t*BA**	*t*-butyl acrylate
TBD	1,5,7-triazabicyclo[4.4.0]dec-5-ene
TEMPO	2,2,6,6-tetramethylpiperidine-N-oxyl
THF	tetrahydrofuran
ToF	time of flight
UV	ultra violet
V_e	elution volume
η	viscosity

Curriculum Vitae

Education

2009 – Present	**Doctorial Studies in Chemistry** Under the supervision of Prof. Dr. Christopher Barner-Kowollik Karlsruhe Institute of Technology (KIT), Karlsruhe, Germany
2008 – 2009	**Diploma in Chemistry** *Macromonomers as versatile synthetic building blocks* Under the supervision of Prof. Dr. Christopher Barner-Kowollik University Karlsruhe (TH), Karlsruhe, Germany
2004 – 2009	**Studies of Chemistry** University Karlsruhe (TH), Karlsruhe, Germany Diploma in Chemistry
2004	**High School Certificate**

Employment History

June 2009 – present	**Institut für Technische Chemie und Polymerchemie** Research group of Prof. Dr. Christopher Barner-Kowollik Karlsruhe Institute of Technology (KIT), Karlsruhe, Germany Scientific Co-worker

List of Publications and Conference Contributions

Refereed Journals Publication List

[5] *Transformation of Macromonomers into Ring-Opening Polymerization Macroinitiators – A Detailed Initiation Efficiency Study*

Zorn, A.-M.; Barner-Kowollik, C. ***J. Polym. Sci., Part A: Polym. Chem.*** **2012**, *50*, 2366–2377.

[4] *Copolymers of 2-Hydroxyethylacrylate and 2-Methoxyethyl Acrylate by Nitroxide Mediated Polymerization: Kinetics, SEC-ESI-MS Analysis and Thermoresponsive Properties*

Hoogenboom, R.; Zorn, A.-M.; Keul, H.; Barner-Kowollik, C.; Moeller, M. ***Polym. Chem.*** **2012**, *3*, 335–342.

[3] *A Detailed Investigation of the Free Radical Copolymerization Behavior of n-Butyl Acrylate Macromonomers*

Zorn, A.-M.; Junkers, T.; Barner-Kowollik, C. ***Macromolecules*** **2011**, *44*, 6691–6700.

[2] *High Temperature Synthesis of Vinyl Terminated Polymers Based on Dendronized Acrylates: A Detailed Product Analysis Study*

Zorn, A.-M.; Malkoch, M.; Carlmark, A.; Barner-Kowollik, C. ***Polym. Chem.*** **2011**, *2*, 1163–1173.

[1] *Synthesis of a Macromonomer Library from High-Temperature Acrylate Polymerization*

Zorn, A.-M.; Junkers, T.; Barner-Kowollik, C. ***Macromol. Rapid Commun.*** **2009**, *30*, 2028–2035.

Conference Contributions

Self Presented Contributions

[2] *'The Free Radical Copolymerization Behavior of n-Butyl Acrylate Macromonomers'*

Zorn, A.-M.; Junkers, T.; Barner-Kowollik, C. *Bayreuth Polymer Symposium* **2011**, Bayreuth, Germany, 11th to 13th of September **2011**, Bock of Abstracts. (*Poster Presentation*)

[1] *'Direct Macromonomer Synthesis via High Temperature Acrylate Polymerization'*

Zorn, A.-M.; Junkers, T.; Barner-Kowollik, C. *Controlled Radical Polymerization Meeting* **2009**, Houffalize, Belgium, 17th to 18th of September **2009**, Book of Abstracts. (*Poster Presentation*)

Contributions as a Co-Author

[3] *'Mass Spectrometry as a Mechanistic Tool'*

Barner-Kowollik, C. *PacifiChem* **2010**, Honolulu, Hawaii, 16th to 20th of December **2010**, Book of Abstracts. (*Oral Presentation*)

[2] *'Efficient Polymer Material Design via Modular Building Blocks'*

Barner-Kowollik, C. *Dresden Polymer Discussion* **2010**, Meissen, Germany, 18th to 21st of April **2010**, Book of Abstracts. (*Oral Presentation*)

[1] *'Direct Radical Appraisal: Complex Reactions in Acrylate Polymerizations'*

Junkers, T.; Koo, S.P.S.; Zorn, A.-M.; Barner-Kowollik, C. *Bunsenkolloqium* **2010**, Göttingen, Germany, 8th of April **2010**. (*Oral Presentation*)

Acknowledgements

First, and foremost, I am indebted to Prof. Dr. Christopher Barner-Kowollik for offering me the possibility to explore exciting projects, the freedom associated with the confidence he has placed in me as well as sustained support during the entire time. A throughout open door and his enthusiasm for science always motivated, even – or particularly – during rough research times.

I also thank Prof. Dr. Thomas Junkers who partly supervised the projects within this thesis until he left for a professorship at the University Hasselt (Belgium). He facilitated getting started in the research area of acrylates.

Many thanks to the entire *macroarc* research team – former as well as present members – for the nice working atmosphere, with a particular thank to the *AC crew* for the strong company and joint enjoyable breaks when necessary in the 'enclave'.

For sharing insights regarding analytical methods I would like to thank Dr. Till Gründling for passing on his knowledge in SEC/ESI-MS and Christina Schmid for the fruitful time on the 2D SEC machine.

In addition, I am grateful for financial support from the *Karlsruhe House of Young Scientists* (KHYS) enabling a research stay at the KTH, Stockholm, Schweden. For the part-time supervision in Stockholm I would like to thank A/Prof. Anna Carlmark and A/Prof. Michael Malkoch as well as the co-workers at the division of Coating Technology for the nice welcome.

Many thanks go to Pia Lang and Tanja Ohmer at the organic department who performed the NMR measurements.

For proof-reading parts of the thesis I would like to thank (in alphabetical order) Dr. Francesca Bennet, Dr. Nathalie Guimard, Dr. Andrew Inglis and Dr. Andrew Vogt. Many thanks also go to Bernhard Schmidt, Kim Öhlenschläger and Dominik Voll for the final trouble-shooting.

Many thanks to Dr. Andrew Inglis and Bernhard Schmidt who bloomed scientific manuscripts with TOC Figures. In addition, I thank Bernhard Schmidt for the cover graphic.

Zu guter Letzt möchte ich meinen Eltern, meinem Bruder, meiner gesamten Familie, meinen Freunden und Jan von Herzen für die anhaltende Unterstützung und die Begleitung auf meinem bisherigen Lebensweg danken.